Civil Wars: A Very Short Introduction

VERY SHORT INTRODUCTIONS are for anyone wanting a stimulating and accessible way into a new subject. They are written by experts, and have been translated into more than 45 different languages.

The series began in 1995, and now covers a wide variety of topics in every discipline. The VSI library currently contains over 750 volumes—a Very Short Introduction to everything from Psychology and Philosophy of Science to American History and Relativity—and continues to grow in every subject area.

Very Short Introductions available now:

Available soon:

For more information visit our website

www.oup.com/vsi/

Monica Duffy Toft

CIVIL WARS

A Very Short Introduction

OXFORD
UNIVERSITY PRESS

OXFORD
UNIVERSITY PRESS

Oxford University Press is a department of the University of Oxford.
It furthers the University's objective of excellence in research, scholarship,
and education by publishing worldwide. Oxford is a registered trade mark of
Oxford University Press in the UK and in certain other countries.

Published in the United States of America by Oxford University Press
198 Madison Avenue, New York, NY 10016, United States of America.

Library of Congress Cataloging-in-Publication Data

Names: Toft, Monica Duffy, 1965- author.
Title: Civil wars : a very short introduction / Monica Duffy Toft.
Description: New York : Oxford University Press, 2024.
Identifiers: LCCN 2024009104 (print) | LCCN 2024009105 (ebook) |
ISBN 9780197575864 (paperback) | ISBN 9780197575888 (epub)
Subjects: LCSH: Civil war—History—20th century. |
Civil war—History—21st century. | War—Causes—20th century. |
War—Causes—21st century. | Conflict management—History—
20th century. | Conflict management—History—21st century.
Classification: LCC JC328.5 .T64 2024 (print) | LCC JC328.5 (ebook) |
DDC 355.02/18—dc23/eng/20240316
LC record available at https://lccn.loc.gov/2024009104
LC ebook record available at https://lccn.loc.gov/2024009105

Integrated Books International, United States of America

Contents

List of illustrations

Introduction

The intricate and multifaceted relationship between humans and
civil wars has profound implications for our comprehension of
violence and its impact on modern society. Civil wars, by
definition, occur within the borders of existing polities or states,
but their repercussions often extend far beyond these boundaries.
They have substantial humanitarian consequences, inflicting
immense suffering. Additionally, civil wars often give rise to illicit
networks involved in the trade of weapons, drugs, and even
human beings. These conflicts raise profound ethical questions
about the use of force and the protection of human rights within
states; questions which continue to inform the development of
international norms and laws governing warfare and military
occupation. Furthermore, in our interconnected world as civil
wars cross national borders, escalate, and encourage foreign
intervention they have far-reaching consequences for global peace,
justice, and prosperity. These interconnections—the variation in
their causes, how they are fought, how they evolve, and how they
end—therefore remain vital.

At their core, civil wars reflect facets of human nature,
encompassing human nobility and self-sacrifice, fear and
insecurity, aggressiveness and greed. Knowledge of how these
dimensions of human behavior is implicated in civil wars is crucial
for recognizing their potential for violence, escalation, and

repetition, which, in turn, enables the development of conflict prevention and mitigation strategies, and effective policy implementation of those strategies.

It is not just human psychology that underpins civil wars. The economic, social, and political contexts in which violence emerges are critical as well. Political decisions, leadership, and governance structures have a significant impact on the escalation or de-escalation of conflicts. Often it is acknowledgment of these factors that is vital for conflict management and peace promotion. Competition for scarce or finite resources too serves as a catalyst for violence and wars. Outlining the connection between resource scarcity and conflict is therefore essential for addressing these challenges and ensuring sustainable resource management. Disparities in wealth and power frequently lead to societal unrest and violence. Acknowledging these inequalities is fundamental to fostering social justice and equitable systems that reduce the potential for conflicts. Finally, the way previous conflicts are remembered—so often subject to manipulation for political, economic, or social purposes—dramatically affects how affected populations come to see the stakes involved in any conflict.

So, it is encouraging to know that although every civil war is unique when we look at the known history of civil wars we see patterns. States susceptible to civil wars share several characteristics and factors. While no single type of state is universally prone to civil war, specific conditions and state attributes can increase the likelihood of civil conflict. These include:

1. *Precedent*: states with a history of unresolved conflicts or a legacy of previous civil wars may be more susceptible to renewed violence, especially if underlying grievances persist.
2. *Fragile institutions*: states with weak governance, limited capacity to provide public services, and a lack of control over their territory

are often more susceptible to civil war. Weak states may struggle to enforce the rule of law or maintain order.

3. *Divisions and grievances*: states with significant divisions and hostility between political and ideological factions may be more vulnerable to civil conflict, particularly when these divisions are associated with historical grievances or discrimination. Such polarization may erode trust among groups in society and in their institutions.

4. *Political repression*: states that suppress political dissent and deny civil liberties may face greater risks of civil conflict. Such regimes can create an environment where grievances fester and opposition becomes radicalized against the state.

5. *Political exclusion and identity politics*: states with political systems that exclude some groups from decision-making processes or hinder their political representation can see these groups turn to violence to assert their rights. Nationalism and identity politics might be ripe for manipulation by elites, furthering divisiveness and radicalization.

6. *Corruption*: corrupt states that undermine trust in government institutions increase the potential for civil unrest, as citizens become inclined to protest or rebel against the regime's corruption.

7. *Economic inequality*: states with high levels of economic inequality or unequal access to key resources, including potable water, arable land, or energy, can lead to social discontent and unrest, potentially escalating to civil war.

8. *Bad location*: states located in dangerous neighborhoods or bordering other states with tumultuous politics are more prone to escalation to civil war. Geopolitical interests may exacerbate instability, or neighboring countries or external actors might insert themselves into a state's internal affairs, escalating tensions and contributing to civil war.

Important to note is that these factors often interact and overlap, and the presence of these risk factors does not guarantee that a state will experience civil war. However, they do heighten the potential for political instability and violence.

Furthermore, even stable democratic states are not immune to the risk of civil war. While democracies are generally characterized by their commitment to the rule of law, inclusive political systems, and respect for individual rights, several factors can make them susceptible to internal conflict. The path to susceptibility in such states may differ from that in fragile states, but the risk remains. Consider these factors:

1. *Prior history of conflict and war*: deep-rooted historical tensions, grievances, or unresolved conflicts can resurface and escalate, especially if there is a lack of reconciliation or acknowledgment of past injustices. Even in democracies, there may be inadequate conflict resolution mechanisms, such as judicial processes or mediation, making it challenging to address disputes peacefully.

2. *Weakening institutions*: in mature states, a gradual erosion of trust in institutions or weakening of governance can undermine political stability. If state institutions fail to adapt to new challenges or address the evolving needs of the population, it can lead to civil unrest. In some democracies, weak governance, corruption, or ineffective institutions can undermine state legitimacy and its ability to address citizens' concerns. A sense of government ineffectiveness can lead to protests or civil unrest.

3. *Ethnic, sectarian, and cultural divisions*: over time, ethnic, sectarian, or cultural divisions within a stable democratic state can become more pronounced, especially if a minority group grows at a faster rate than a majority group. These divisions can evolve into grievances or conflicts if they are not properly addressed, potentially leading to ethnic or identity-based violence. Changes in leadership and demographics within mature states can influence this susceptibility. A younger generation representing the identity group emerges with different expectations and sets of grievances, challenging the political status quo and potentially fueling civil unrest.

4. *Rejection of majority rule and radicalization*: the emergence of radical or extremist movements within these states can pose a significant risk. Such movements may exploit grievances or identity politics to mobilize followers and incite violence. Democracies can face threats from dissatisfied groups that reject the democratic system and use violence to advance their agendas. These groups may be difficult to contain and could incite broader conflict.

5. *Political polarization*: long-standing political stability can erode if political polarization emerges and deepens. In mature democracies, intense partisan divisions and a loss of faith in the political system can escalate tensions and create an environment where conflict becomes a real possibility. Gridlock, distrust, and the delegitimization of opponents arise, and in such cases non-violent conflict resolution processes—a tradition of debate and compromise—begin to wither, pushing factions toward conflict as they seek to secure their interests and beliefs without compromise. The rise of nationalist or populist movements can stoke divisions within democratic societies, emphasizing "us versus them" narratives that can escalate into civil conflicts when critics of incumbent policies are rebranded as "enemies of the state" and an elite emerges claiming that they alone can solve the ills of the country.

6. *Economic disparities and shocks*: high levels of economic inequality can undermine the stability of democratic states. When a significant portion of the population perceives its economic prospects as bleak or worsening, it may resort to protests or even violence as a means of redress. Even in these states, severe economic downturns, such as a financial crisis or recession, can lead to widespread economic hardship and inequality. Competition for finite resources, especially in regions prone to environmental or climatic challenges, can exacerbate societal tensions. The resulting social discontent can create fertile ground for civil unrest and conflict, even in countries with a history of stability.

7. *External factors*: external influences, such as regional instability or geopolitical maneuvering, can impact the stability of a stable, democratic state. Proxy conflicts and the involvement of external actors can exacerbate internal tensions and increase the risk of civil conflict. External actors, including neighboring countries, may exploit political, ethnic, or religious divisions within a democratic state to further their own interests. This external meddling can contribute to internal instability and potentially larger-scale violence and civil war.

8. *Social media accelerated mis- and disinformation*: in the digital age, the rapid spread of misinformation and disinformation on social media can exacerbate existing divisions, fuel mistrust, and incite violence. It can create the false impression that a fringe idea with very few adherents is in fact a mass movement worthy of inclusion in political, economic, or social policy discussions. Although information flow and spread is not new, the intensity and speed of it may undermine the politics of even the most stable democratic states.

The transition from a stable democratic state to susceptibility to civil war is rarely a linear process; it can result from a combination of factors that weaken the state's stability and social cohesion.

Furthmore, unlike more traditional wars, the causes of civil wars and differences in the way they are fought and end are often long term, indirect, and intangible. In terms of complexity, consider that there are only 193 member states in the United Nations, making over 18,000 possible conflict dyads. By contrast, there are over 10,000 nations bound by these states, so if we only look at civil wars in which nationality is a factor, that number of potential conflict dyads expands to tens of millions. And so many civil wars—such as the long fight between Irish Protestants and Irish Catholics in Northern Ireland—engage *multiple* cross-cutting issues beyond nationality, such as religious faith, social class, and claims to homeland territory. Civil wars are much more likely than interstate wars to involve insurgents that do not wear uniforms or follow a single authority and given that many tend to last a long time, allegiances and grievances often change and shift.

This complexity and nonlinearity underscore the importance of continuous efforts to address discontent, promote inclusive governance, and adapt to evolving challenges to maintain long-term stability. Recognizing these potential risk factors and working to mitigate them is crucial for preventing the onset of civil conflict in even the most stable states.

This Very Short Introduction to the factors influencing civil wars and their impact on various types of states highlights the relevance and value of this knowledge for a broad audience. It is essential for the general public, academics, policy-makers, humanitarian workers, journalists, and business leaders alike. This book fosters global awareness of critical factors, empathy for those suffering from destructiveness and displacement of these wars, and a sense of responsibility for issues related to that violence and humanitarian crises.

For scholars and students of history, political science, international relations, or conflict studies, this book offers a more holistic sense of the complexities surrounding civil wars.

Policy-makers and diplomats can learn about the causes and consequences of civil wars and consider how this knowledge might shape their approach to national and international policies.

Professionals working in conflict zones or with displaced populations can benefit from insights into the origins and endings of civil wars and the challenges faced by civilians because of them, potentially aiding their efforts to provide more contextualized—and therefore more effective—aid and support to affected communities. Individuals and organizations involved in peacebuilding and conflict mediation can draw from the lessons and experiences documented within these pages to inform their strategies and interventions.

Those engaged in human rights advocacy, peace movements, and international development can use the insights from the scholarship featured here to bolster their arguments and campaigns, driving awareness and change.

Journalists covering conflict zones can enhance their reporting by deepening their knowledge of the underlying causes and context of civil wars, leading to better informed news coverage.

Finally, business leaders operating in regions affected by civil wars can benefit from awareness of the security, political, and economic factors at play, which is essential for making informed business decisions and managing risks.

While this book is concise, and therefore cannot be comprehensive, it nevertheless provides historical context, critical insights, and an appreciation of the complex nature of civil wars, hopefully contributing to informed decision-making, reduced human suffering, and a more peaceful and stable world.

Chapter 1
A short history of civil wars, and why civil wars matter

Civil wars are nasty, brutish, and on average, far from short. They tend to last for years and display a destructive tendency to escalate and to expand geographically. The modern state of Sudan, for example, had known only thirteen years of peace since it gained independence in 1955 until 2005. Afghanistan has been riven by armed conflict almost continuously since 1978; and what began as a civil war in the Democratic Republic of Congo has been called Africa's first world war because its vortex of violence drew in so many of its neighbors.

What we know of civil wars, and how we know it, has changed a great deal over the past century. Prior to the Second World War, the study of civil wars was given over to historians and, in the social sciences, to those studying comparative politics. The study of civil wars was often accompanied by years of field work, interviews, archival research, and the study of foreign languages. Civil wars are, in general, much more complex than interstate wars. There are many more political actors (nations, for example, number in the tens of thousands globally; whereas there are fewer than 200 recognized states), and the motives for mobilization and escalation tend to be unique to each conflict. This lent inquiry into interstate and nuclear warfare—both marked by a tendency to concentrate destruction into fewer years as compared to civil and ethnic conflicts—an air of superiority. This is especially evident since the publication of Kenneth N. Waltz's *Theory of International Politics* (1979), which compelled international relations scholars to emulate the natural sciences in creating general theories.

Since the end of the Cold War, when the number of civil wars seemed to be on the rise (they were not), the study of civil wars has shifted to those who pursue international relations (and its variants) as subfields of political inquiry. The number of scholars who undertake field work or struggle to master foreign languages has dropped as compared with those whose work relies on sophisticated statistical analyses or formal models. Again, in a global North context, this mechanization of inquiry appeared to offer a more objective and more "scientific" appreciation of the causes, nature, and termination of civil wars.

The arid quality of mechanized inquiry also led, arguably, to mitigation strategies and policies that proved unworkable or counterproductive. These tended, unsurprisingly, to be based on things that could be counted and compared such as corpses, lootable resources such as gold, oil, or diamonds, and so on. What remains lamentable about the shift is not that mechanized inquiry yields little of value to our knowledge of civil wars, but that its value remains dependent on an older and more intimate knowledge of human geography; of identity and the many ways humans bring meaning to objects.

Civil wars in history

There is no space in a very short introduction for anything approaching a full listing or summary of early civil wars, so I have selected from those most representative, as best we can tell, of civil wars as we know them today.

Across the sweep of history, from ancient civilizations to the complexities of modern geopolitics, civil wars have played a defining if at times underestimated role. These conflicts are characterized by their destructive impact and relentless, often indiscriminate violence; as well as enduring scars and remembered grievances.

Civil wars in antiquity still hold powerful lessons for us today, but as always with history, we must remember that how "we" approach an historical event is bound to be as much about what we bring with us to the facts, as any supposedly objective fact itself. Anachronism, in other words, remains a persistent threat to knowledge.

Perhaps the most important difference between how we view armed conflict today and how it has been viewed in many places in the world and in many times is the idea of the "non-combatant." From the Middle Ages to the present, we can say with greater confidence that every human society recognized a category of human community that was off limits to intentional injury (e.g. "women and children"). But before that the record becomes more fraught.

With these cautions in mind, one might argue that the first recorded history of a civil war took place in ancient Mesopotamia around the twenty-first century BCE. The Ur dynasty (also known as the Third Dynasty of Ur) ruled in the region of Sumer, which lies in contemporary Iraq. Once considered the most powerful and prosperous in Mesopotamian history, the Ur dynasty eventually collapsed. Among the factors we can identify today that led to its collapse is internal dissent, as different factions descended from its ruling class began to struggle for control over the capital city of Ur and its rich adjoining territories. As the center weakened, the broader region fell prey to foreign invaders, leading to a period of political fragmentation as competing city-states and dynasties fought for power and control.

Internal disputes rising to the level of civil war also followed the death of Alexander the Great in 323 BCE. Alexander's empire stretched from Greece to India. Under his rule, imperial lands were divided among his top generals, known as Diadochi or "Successors." These generals had been appointed by Alexander as regional rulers, and after his death they fell into dispute over the relative value of their regions as compared with their peers. The escalation of these disputes into the Wars of the Diadochi resulted in overlapping military campaigns, the formation then breaking of alliances, and treachery and intrigue. Although not a textbook case of civil war because Alexander's empire was not a *state*, the nature of the empire's disintegration and descent into inter-regional violence parallels the all-too familiar pattern of civil war as we know it today.

Turning to East Asia, we can identify many of these same patterns and motivations in China's Three Kingdoms period (184–208 CE), following the fragmentation of the Han dynasty (420–589 CE) into three competing kingdoms: Wei, She, and Wu. Later reunified during the Jin dynasty of 420–589 CE, that dynasty would itself disintegrate into warring dynasties that achieved brief periods of consolidation, before again being torn into constituent and warring parts. This process—resembling in every way, but not exactly matching, the pattern of a contemporary civil war—did not end until 1913, when the fall of the Qing dynasty led to the establishment of the Republic of China. The Republic of China was itself soon overwhelmed by civil war, this time between Chinese Communists and Nationalists. That civil war only ended in 1949, with the formation of the Peoples Republic of China and the flight of the Nationalists to Formosa (now Taiwan).

Other records of civil war, or their close analogs, can be recovered in the Old and New Testaments of the Bible. These biblical accounts illustrate internal strife and political division rising to the level of violence. Because these texts are not considered to be literal descriptions of historical events so much as allegories based on historical events, we need to approach them with caution. That said, many of these biblical allegories contain descriptions of strife and violence we would today recognize as civil war. In the Books of Judges and the prophetic writings, for example, there are many accounts of rebellions leading to internal conflict among the Israelites and their leaders. The Rebellion of Absalom (ca. 1000–970 BCE) referred to in 2 Samuel 15–18 tells of a conflict that emerged after King David's son Absalom rebelled against his father, throwing the kingdom of Israel into violent conflict. Absalom was later killed in the forest of Ephraim.

Likewise, King Solomon's reign, as recounted in 1 Kings 12, suffered familial strife when his son Rehoboam, faced a rebellion led by Jeroboam who, around 930 BCE, sought to create a "northern" kingdom of Israel. That rebellion ended with the kingdom divided into a northern kingdom of Israel and a southern kingdom of Judah, each with its own ruler.

Just as in the Judeo-Christian texts, Muslim religious texts also contain records of internecine violence we would today recognize as civil war. Such narratives are mainly found in the Hadiths, or "sayings of the Prophet", which along with the Quran form the core of Islamic religious practice.

Among the most significant conflicts recounted in the Hadiths is the Battle of Badr (624 CE), which pits a Muslim community in Medina against the Quraysh tribe of Mecca. Other important battles between Muslims and the Quraysh included the Battle of Ehud (625 CE), the Battle of the Trench (Ghzawat al-Khandaq, 627 CE)—in which a coalition of Arab tribes besieged Medina— and the Conquest of Mecca (630 CE), which was seen as a pivotal event in early Islamic history. Keeping in mind that scholars of civil war rely on the idea of a state *within which* wars engage rival groups, and the "state" as we know it today did not exist in the seventh century, what we take from these battles is their similarities to contemporary civil wars. For example, not only was the conquest of Mecca taken to symbolize the triumph of Islam (and God's favor), but it fulfilled prophetic promises, promoted forgiveness and reconciliation, and consolidated religious and political authority. Most post-seventeenth-century civil wars share these qualities, including the deep-seated divisions that persisted after fighting ceased. As we will see in later chapters, within the Muslim world after 630 CE these remembered divisions have served as pretexts for civil wars in contemporary times (e.g. Iraq following the US-led invasion of 2003).

South Asia is also home to a rich history of war and peace, often allegorical rather than literal, but serving the purpose of instruction in secular and religious traditions. Extending from the third century BCE to the fourth century CE, for example, the epic *Mahabharata* depicts a dynastic struggle for power between cousins, who were locked in a battle for the throne of Hastinapura. It is difficult in scale and artistic quality (the poem in its final form stands at over 1.8 million words) to identify a comparable epic centered around a famous dynastic war (the Kurukshetra War, date uncertain). So although we would recognize the conflict recounted in this sacred and foundational text in Indian culture as a civil war in its dynamics, as in other non-historical sources in analogous Judeo-Christian or Muslim texts, the historical facts on which the epic appears to be based remain elusive.

Perhaps one of the most famous "civil wars" for which we have factual historical records is England's "War of the Roses" (1455–1487), between the noble houses of Lancaster and York— each a component of the Plantagenet dynasty. Each house had a legitimate claim to the English throne but considered possession of the throne to be an indivisible issue. That zero-sum construction of the stakes of the conflict—very common in civil wars more broadly—likely motivated the more than three decades of fighting that only ended with a dynastic marriage that combined the two houses into one: the Tudor dynasty.

Another dynastic civil war—and in terms of the history of contemporary international relations theory, perhaps the most important—began as a dynastic civil war among the Hapsburgs in 1618. This war would expand and escalate into a religious war lasting 30 years, depopulating much of Europe. The Treaties of Westphalia, which ended the Thirty Years' War, established the principle of *sovereignty*, or the notion that rulers could do as they please within the territories they governed, and interference from others was no longer accepted as legitimate. This gave rise, over the following decades, to what we take for granted today: the states system.

Of course, dynastic conflict, such as those described above, are only one subset of civil wars more broadly. And while dynastic disputes typically entail rival factions related by blood ties vying for leadership or control, many do not rise to the level of civil war as we think of it today. In fact, it is worth underlining here that a war typically involves (1) organized violence between at least two sides, each of which possesses (2) some capacity to seriously injure the other, and which results in (3) at least a thousand battle deaths in a calendar year. And it is also worth remembering that in places and times in which people believed that virtues were entirely hereditary and victory in fighting was evidence of a god's favor, blood ties had a different significance than they would today, after Darwin and after the advent of the science of genetics.

In Europe what we witness is a slow shift away from the idea that one is born with all the capacities one needs to rule justly and effectively, and that the Christian God's favor necessarily entitles children to follow fathers as rulers. The very idea of the centrality of monarchy came under attack in one of the earliest and most devastating civil wars in the post-medieval period, the English Civil War (1642–1651).

That war pitted supporters of King Charles I against a group of aggrieved parliamentarians, who sought greater power at the expense of an absolute monarch. Bitter disagreements over taxation and the extent of royal authority, as well as religious identities, escalated to armed conflict. The English Civil War marked a major turning point in English history. It resulted in the temporary overthrow of the monarchy, the trial and execution of Charles I (1649), the establishment of the English Commonwealth, and later a Protectorate under the rule of Oliver Cromwell. As a classic civil war, the English Civil War featured regular armies under organized command authority, motivated by opposing conceptions of legitimacy (zero-sum)—particularly regarding matters of faith—and marked by a cruel and devastating impact on what we would today call non-combatants.

A little over hundred years later, another pivotal civil war would engulf the North American continent in 1861: the US Civil War (1861–1865). The war pitted a group of largely agricultural Southern states (the Confederacy) against a largely industrial group of Northern states (the Union). It tested the very founding principles of the nation, ultimately preserving the Union but at tremendous cost. It is considered the first modern industrial war and a quintessential civil war, with distinct sides, each commanding organized armies and engaging in large-scale battles. The ferocity of this fight, its scale, and its use of technology in warfare differentiated it from earlier conflicts—as did the use of mobile artillery, support by railroad, and repeating rifles, as well as ironclad warships with steam engines that heralded the obsolescence of traditional wooden warships. US general William Tecumseh Sherman also famously attacked civilian infrastructure—burning houses and slaughtering cattle—in his infamous "march through Georgia," a practice tragically common in civil wars then and now.

The bloodiest civil war for which records survive took place in China and lasted from 1850 to 1864. The Taiping Civil War (also known as the Taiping Rebellion) resulted in between 20 and 30 million deaths, or anywhere from one-tenth to one-twentieth of China's pre-war population. As with many civil wars, especially those involving religious motivations, as this one did, civilians were frequently victims as well.

Following the Russian Revolution and the fall of the Russian Empire, various factions, including the Bolsheviks (Reds) and anti-Bolshevik groups (Whites), engaged in a protracted struggle for control of Russia (1917–1923). The Bolsheviks prevailed, resulting in the end of the Romanov dynasty and the establishment of the Union of Soviet Socialist Republics (USSR or Soviet Union), a state nominally governed by communist ideology for nearly seven decades. What is striking about this war was that it exhibited both traditional forms of fighting that included irregular troops and partisan warfare with fluid frontlines and a host of local and regional conflicts in conjunction with industrial technology, including machine guns, artillery and armored trains, and reliance on railways and telegraph communications to deal with the long distances needed to manage the large-scale logistics and supply lines. It also proved disproportionately hard on civilians, who were often deliberately targeted by opposing forces.

Like the civil war in Russia, twentieth-century Spain witnessed a deep ideological divide that tore the nation apart. The Spanish Civil War (1936–1939) served as a precursor to the larger ideological battles of the coming (second) world war and typified how ideological divisions can drive a people apart. Republicans and Nationalists, motivated by complex ideological conflicts involving fascism, communism, and anarchism, engaged in a brutal struggle. The Republican side represented a coalition of leftist, liberal, and socialist forces, while the Nationalist side comprised right-wing, conservative, and fascist adherents. It also marked the transition from more traditional civil wars over identity such as religion and race to those with ideological dimensions and international implications; and it featured a significant foreign military intervention as Germany's Third Reich sent combat forces to aid anti-Republican forces, and volunteers from the United States and other countries traveled to Spain to fight and die on behalf of the Republicans.

All these civil conflicts, spanning millennia and continents, share common threads. Like modern civil wars, the Ur III dynasty faced internal dissent. Local rebellions and opposition to central authority were early signs of turmoil. Similar to some modern civil wars, the ancient war of the Diadochi centered on the struggle for power and control over territory. The *Mahabharata*, though a work of myth and legend, carries timeless themes of dynastic conflict and justice and continues to inform Indian culture and ideas. Considerable insights may be gleaned from its review.

What is more striking, however, is that contemporary scholars rarely consider these earlier cases in their analyses, and when they do they tend to treat them as one-off episodes that offer few insights into other civil wars. When examining civil wars, particularly from a comparative and especially from a statistical perspective, scholars and analysts typically start with those that began after 1945.

This last point highlights important aspects of academic research and analysis: the focus on contemporary cases and limitations of historical comparisons. Beyond the problem of anachronism noted above, there are several other reasons scholars might not delve into earlier historical cases. First, there are limitations on data availability and their reliability. Again, consider the history of the Ur dynasty, which was recorded on clay tablets. Not only is the reliability of the historical record in question, but standardized data are lacking. A second limitation has to do with relevance. Does the style of fighting with limited technology or over dynastic pursuits relate readily to wars of ethnic nationalism and territorial disputes? Can the same lessons be learned? A third limitation is that scholars often specialize in specific time periods and/or regions of the world such that their research interests may not encompass earlier historical cases.

Fourth, and perhaps more consequently, contemporary theories attempting to illuminate the causes, nature, and termination of civil wars have been adapted from both the international relations and comparative politics sub-disciplines of political science. As noted earlier, a growing dependency on machine-readable "data" may cause research to suffer because even though there are now many databases cataloging hundreds of cases of civil wars and their contributing factors none of them go back before the twentieth century.

Why civil wars matter today

Both the contemporary study of civil wars and the very nature of the wars themselves have evolved since 1945. As observed above, the study of civil wars has changed from a tradition of comparison of historical case studies to sophisticated statistical analyses and formal modeling. Much of this is welcome, but an over-reliance on statistical data analyses has also harmed our knowledge of exactly what is happening in specific places. Put simply, civil wars remain too complicated to be adequately captured by reducing analysis to binary variables. It is not enough to say a war is about greed or fear, providing a coding of "0" or "1" in order to capture why a war emerged or progressed the way it did.

Being cognizant of the historical and human dimensions remains every bit as critical. Greed for what? How intense a fear, and of what? All wars are visceral and emotional, and they result in killing and the destruction of people and their communities— communities that at one time co-existed but somehow exploded into violence. The violence consumes everyone from combatants to non-combatants and often entails acts of terrorism and crimes against humanity, including ethnic cleansing, systematic rape, and genocide. And the memory of barbarism—sometimes discounted but often intensified through political manipulation—is likely to be the prime motivation for subsequent re-ignition of violence.

After Josip Broz Tito's death in 1980, for example, Serb violence against Bosnians a decade later was powerfully motivated by recovered and politically manipulated historical grievances over the defeat of Serbian prince Lazar Hrebeljanović by Ottoman Sultan Murad I in 1389, a conflict that happened hundreds of years prior. In contemporary civil wars, such as that which engulfed Yugoslavia in the 1990s, combatants often know one another personally as killing, raping, and looting move from home to home and neighborhood to neighborhood. Ending such wars often means undertaking the nearly impossible task of setting aside trauma and grievances to try to come together and co-exist again. Neighbors that savaged each other must somehow live together peacefully once more.

In sum, civil wars—still and always nasty, brutish, and apt to escalate, endure, and re-ignite—remain both critical and under-researched. The advent of the Internet and the accelerating global distribution of smart phones has qualitatively altered the ability of disgruntled minorities to organize and to recruit. Mere rumors, accelerated by social media in exchange for advertising revenue, have already caused lethal riots in India and remain a serious challenge to stable governance worldwide. So, in the absence of a greater appreciation for the importance of civil wars, the increase in civil conflict we observe today may only be the vanguard of a more dangerous and violent world tomorrow.

To capture the broader trends in civil wars today, a good place to begin is with the Peace Research Institute Oslo (PRIO), which has been working for decades with the Uppsala Conflict Data Program to count and catalog instances of political violence, including civil wars. For example, its 2020 report underlines some distressing trends. First, in 2019, the number of civil wars reached an all-time high since 1946, the start of their collection effort. It recorded 54 state-based conflicts, defined as contests involving a state over control of the government or some piece of territory with at least

25 battle deaths within a calendar year, with the deadliest civil wars occurring in Afghanistan and Syria.

Data to 2022 show quite clearly that civil wars and their internationalized counterparts are the main form of armed conflict worldwide, with interstate wars becoming quite rare and colonial wars disappearing since the mid-1970s (Figure 1).

Three other trends stand out. First, the number of conflicts peaked in the early 1990s after the end of the Cold War. The collapse of the Soviet Union and the emergence of war in places like Chechnya, Moldova, and Tajikistan contributed to the rising numbers. Second, the number of wars soon fell as the 1990s progressed. The decline was due to the end of the Cold War rivalry that served to underpin proxy wars in the 1970s and 1980s, with resources and support for warring factions drying up. This soon led to an outbreak of peace agreements around the world, including in Bosnia and Herzegovina, Timor Leste, El Salvador, Sierra Leone, and Tajikistan. Third, the number of battle deaths is declining. This is good news, although the number is still hovering around 50,000 a year, and "battle deaths" remains a limited way of describing the experienced destruction of war. Only a few countries account for a majority of battle deaths, most notably Afghanistan and Syria. In 2018 and 2019 Syria lost about 1.6 percent of its population and Afghanistan lost almost 1 percent of its own (Figure 2).

Clearly, civil wars are deadly and destructive, and all too common. In fact, today civil wars are the deadliest form of political violence. Moreover, although civil wars tend to occur in poorer, less developed countries, wealthier and more developed (both politically and technologically) societies are far from immune. When I was first starting my own research into civil wars, I remember a particularly distressing account from a colleague in anthropology who had been studying the contested territory of Nagorno Karabakh. Outright violence broke out between

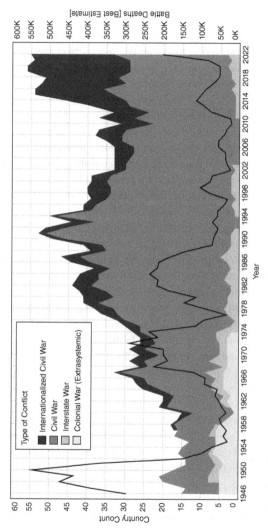

1. Number of countries with state-based armed conflict and war trends: 1946–2022

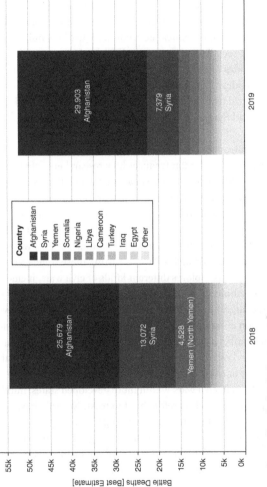

2. **Share of battle deaths in 2018 and 2019**

Armenians and Azerbaijanis in 1988, and it remained a hotly contested zone until 2024 when Nagorno Karabakh/Artsakh was dissolved following a successful September 2023 Azerbaijani offensive that caused ethnic Armenians to flee. Although prior to 1988 Nagorno Karabakh had been a highly developed region, my colleague recounted just how quickly it de-civilized socially and economically as residents were forced to choose sides and face living without electricity and running water in modern high rises. The same holds true for regions of former Yugoslavia, the city of Grozny in Chechnya, as well as Lebanon and Syria.

This book introduces how civil wars emerge and the consequences when they do. Chapter 2 introduces how researchers define civil war and the thresholds of battle deaths on which they commonly rely. It discusses how civil wars compare to riots, protests, and non-violent forms of civil resistance, and how they relate to different forms of political violence like terrorism and mass killing of civilians. It reviews how often and how intensely civil wars have been fought since World War II, contrasting these trends with interstate wars. It explores different kinds of civil wars, including territorial, ethnic, religious, and nonreligious conflicts. Morbidly, one of the largest debates in the formal study of civil wars is over the question of how many people must die before we can consider a violent conflict a civil war? 25? 100? 1,000? Because civil wars are lethal and often happen in underdeveloped places with little government or outside presence, obtaining accurate counts is difficult. Even in civil wars within more developed states, numbers can be politicized and therefore hard to trust. Nevertheless, scholars scour the best available sourcing they can from journalists, nongovernmental agencies, religious organizations, and locals as part of their effort to record the history of the fight and its attendant death and destruction.

Chapter 3 introduces and explores the risk factors that make a country prone to civil war and different theories analysts have crafted to explain why a civil war might emerge. Prominent theories include identity issues like ethnicity and nationalism, economic grievances and greed, rationalist explanations like the security dilemma and commitment problems, both of which are based in fear, and manipulation by elites. As noted above, it is intriguing how the staying power of old grievances (things like a lack of political access or economic complaints) keep tensions high and the emergence of other issues (notably religious identity and different interpretations for how religion should play out publicly) become more central to the conflict.

Once a civil war begins, there are both immediate and long-lasting effects on citizens' lives and livelihoods and a country's economy, governing structures, and stability. Chapter 4 considers the immediate effects on citizens, including deaths, displacement, disease outbreaks, and loss of employment and education, and then analyzes the long-term economic, political, and security implications of civil wars, particularly the risk of war recurrence. When we try to understand the real costs of war, it is important to consider a few "what if" questions. What if the war had not happened? What if the United Nations had not intervened? What if the country had been divided (partitioned) into two new states? Or, what if, having been partitioned, it had not been?

The answers to these questions do not matter only to the country at war. The effects of civil wars often, and increasingly, spill over international borders. Refugees flee to other countries, warring parties might operate in relative safe havens across international boundaries, conflicts increase human trafficking and ease illicit trade in arms, drugs, and other illegal goods, and armed groups can potentially mount transnational terrorist attacks. Chapter 5 examines how civil wars affect the economies and security of neighboring countries and the wider regional setting.

Outsiders often get involved in civil wars either to help one side win, to take advantage of a country amid a civil war or revolution (think of Iraq's attack on Iran in 1980, for example), or to establish or re-establish negotiated peace agreements. Chapter 6 outlines the increase in foreign military interventions in recent civil wars and the experiences of United Nations and other external peacekeepers as it has evolved since the 1990s. The effectiveness of peacekeeping has been a hotly debated issue. Does it help or hinder conflict resolution? It also assesses the evolving international laws and norms shaping outside involvement, including the responsibility to protect (R2P) doctrine, which aims to obligate member states of the United Nations to take steps to protect vulnerable people to end their persecution and the worst forms of violence, specifically mass atrocity crimes (genocide, ethnic cleansing, war crimes and crimes against humanity) even if it means the steps taken violate the sovereignty of the target state. Tragically, mass atrocities remain a common feature of civil war.

The final chapter discusses the different ways civil wars end— peace deals or government or rebel victories—and the durability of these different endings. It looks at the history of conflict termination and the likelihood of war recurrence. Controversies over which types of endings should be pursued by outside powers continue, with some people making the case "to let the wars burn" in a Darwinian nod, while others advocate for mediation and negotiations even if evidence of their lack of effectiveness might suggest a different policy. The chapter concludes with some thoughts about what the future of civil wars might hold.

Although there is a lot we still don't know, over the past 30 years our knowledge of civil wars—their origins, consequences, and conclusions—has broadened and deepened significantly. This is in part because like the philosopher Thomas Hobbes whose worldview was significantly shaped by the English Civil war, the times in which a researcher lives and works tend to attract them to some questions over others. Both the theories researchers use and the evidence they choose to select from history to support their theories also follow. After the end of the Cold War in 1991, civil wars seemed to be everywhere, yet inertia from overriding fear of a global thermonuclear war between superpowers meant we had relatively few scholars whose work on civil wars was commonly sought out or known. Today the study of civil wars has become robust and may give us important insights into how to constructively resolve so many of the problems that appear to polarize us, such as climate change, faith in politics, refugees, immigration, military and humanitarian intervention, and peacekeeping.

Chapter 2
What is a civil war?

Let me start with a brief story. In November 2012, over a year
after massive civilian demonstrations rocked Syria, sparking an
armed conflict and brutal government crackdown, Syrian
president Bashar al-Assad made headlines when he denied in a
news interview that his country was in the midst of a civil war.
Instead, he blamed foreign support and terrorism for
destabilizing Syria). In making his case, al-Assad affirmed, "You
have divisions, but division does not mean civil war. It is
completely different. . . . The problem is not between me and my
people."

International critics were quick to pounce on his assertion. By
mid-2012, the crisis was widely considered to be a civil war, as
most commonly defined by both experts and non-experts.
Syrians took to the streets *en masse* in late 2010 and early 2011
to protest their government's corruption and authoritarianism.
By September 2011, armed opposition groups emerged and
began to stage effective attacks against Syrian forces, but the
government was hitting back just as hard. So when did the
conflict become a fully fledged civil war? When should we say it
actually started?

The International Committee of the Red Cross—widely
considered the "guardian" of the Geneva Conventions, or the

laws that regulate the treatment of non-combatants in war—formally declared the conflict a civil war in July 2012 (prior to Assad's statement) after dozens of deaths were reported in a village attacked by Syrian government troops. The United Nations Under-Secretary-General for Peacekeeping Operations also said so, but a month earlier, when Assad's regime lost control over territory. So who was right? Did the civil war start in May, July, or had it not even begun in November, as Assad argued?

In truth, anyone of them could have been right. No answer is indisputable. The definition of civil wars within the academy and in policy circles is hotly contested; and the definitions vary between policy-makers and researchers. In this chapter, we outline the bases of these arguments. First, we will discuss the definition of civil war, and then contrast it to other similar, but not identical, interstate conflicts. Finally, we will address a point of scholarly contention—whether the nature of civil wars has changed over time. But first, we must consider its basic definition.

Defining "civil war"

It should come as no surprise that the definition of "civil war" is hard to pin down. In their important study, *Resort to Arms*, political scientists Melvin Small and J. David Singer defined a civil war as "any armed conflict that involves (a) military action internal to the metropole, (b) the active participation of the national government, and (c) effective resistance by both sides." Civil wars were distinguished from international or interstate wars by the fact that the violence occurred within the territory of a sovereign state, and that the government participated as a combatant. Both of these characteristics were true in early days of Syria's rebellion as well. Therefore, it is fair to ask: is 2010 when the Syrian conflict turned from a civil resistance movement to a civil war?

Many scholars, including Small and Singer, answer this question using two thresholds, including the intensity of violence and the number of deaths. While both measurements are acceptable at face value, they predictably lead to even more questions. It is unclear, for example, what threshold is needed to tip civic protests and demonstrations that might involve the deaths of citizens and government agents into the civil war category. The bottom line is that numbers can be deceiving, and at some point, *any* number we choose to say "this is" or "this is not" a civil war will be arbitrary. As we will see, however, when it comes to building general explanations, which academics call "theories," some choices work better than others.

Battle deaths

Probably the largest debate around the definition of civil war is about how many people die in battles, or "battle deaths." Small and Singer first argued that the number of battle deaths needed to categorize a conflict as a civil war is 1,000 *per year*, and this number was originally backed up by many scholars, such as in the well-known "Correlates of War" project. Over time, the accepted definition has become 1,000 *total* deaths, rather than 1,000 per year. However, others still argue that both thresholds are too high (similar debates occur over what counts, in terms of those killed in domestic violence, as a "mass killing event"). The Uppsala Conflict Data Program/Peace Research Institute Oslo (PRIO) Armed Conflict Dataset uses a much lower threshold—25 annual battle deaths—while still others advocate for a range rather than a set number. There are also charges that a battle deaths threshold ignores contextual factors. For example, this definition applies equally to a country like the Maldives with a population of 500,000 as it does to India with well over a billion citizens—a population 2,500 times the island nation's size. The Maldives would have to suffer disproportionately to be considered as fighting an actual civil war.

Those who favor a lower body count threshold may have a terrific point, because as humans, wherever we live, we understand that a dead body can mean many different things. Moreover, a simple count of the dead and wounded is unlikely to capture the lived impact or cost of violence. This is why combat veterans returning from war so often envy their fallen comrades, because although they themselves are still breathing, they suffer terribly from the experience of combat, some to the point where they cannot work or love. And feminist international relations theorists have often noted that a focus on body counts as a cost of war often results in discounting the impact of sexual violence in war, such as rape. Finally, although most people believe that the crime of genocide requires that thousands or millions be killed, the Genocide Convention relies not on the *number* of people killed, but on the *intent* to destroy a race, ethnicity, or faith, either through killing or through forcing people from their lands. So it is possible to commit genocide by killing, or even only attempting to kill, a single person.

The 1,000-deaths threshold leads to additional questions. A conflict may be considered a civil war once it reaches 1,000 deaths, but how do we know when to start counting? Does a civil war start the year a conflict begins, even if death totals are lower than 1,000, or does it only officially begin when the death total passes 1,000? Similarly, if we accept that a conflict becomes a civil war when the death count passes 1,000, how do we know when it ends? Some scholars have postulated that a civil war ends when the intensity lowers to less than 100 deaths per year, but there have been few other conclusive answers to this question. As noted above, to some extent any threshold we choose will be arbitrary—there is no real difference in a conflict when the hundredth person dies rather than ninety-ninth.

Lastly, we must ask who counts? Most definitions only consider people bearing arms—combatants—but it is worth wondering whether this is adequate or acceptable. Civilians are often targeted in civil war and are disproportionately affected by the fighting and resulting humanitarian disasters. As Milton Leitenberg once put it: "There were few 'battle deaths' in Cambodia between 1975 and 1978, comparatively few in Somalia in 1990 and 1991, or in Rwanda in 1994: but it would simply be bizarre if two million dead in Cambodia, 350,000 in Somalia and 8,000 or more in Rwanda were omitted from compilations." However, including them in the definition of civil war risks blurring the lines between other types of conflict, such as terrorism, mass killings, and genocide—though these atrocities can definitely occur during civil wars, making the definitions and distinctions even harder to agree on.

Level of violence and organization

Another problem with using the battle deaths threshold, as Leitenberg's quote calls attention to, is that it does not account for the intensity of violence, especially over time. If we were just adding up battle deaths, the severity of conflicts would look a lot less than it actually did on the ground during particularly violent periods. Several scholars attempt to address this by combining a cumulative-death threshold with the requirement that military activity be sustained for the duration of the conflict; however, the definition of "sustained" in this case is still open to interpretation.

Similarly, there is the ongoing question about the required level of resistance and organization on both sides needed to define a conflict as a civil war. There must be an organized opposition group, rather than just an accumulation of killings in a crime-ridden city or state. The Correlates of War project measures effective resistance as the ratio of fatalities of the weaker to stronger forces, which is why one-sided uses of force, such as genocides, do not officially count as civil wars in all cases. But even then, there is no consensus as to what qualifies. Some say that the stronger side must suffer 5 percent of deaths, while others just use the whole number of 100 deaths. There are questions about the time span here, too. Do the state deaths need to occur per year, or can it be over the full duration of the conflict? The answers to those questions have yet to be resolved.

In sum, there are very reasonable arguments on all sides as to what should count as a civil war (and what counts as "war" too), but there is also a lot of disagreement when it comes to picking a universal definition: that simply does not exist. Again, those studying and recording civil wars tend to develop what we think of as a "gut feeling" about the fit between the definitions they chose, and the theories and policies their research supports. It is also worth adding that when policy relies on theory and doesn't work, many would consider that to be a good argument for revisiting how the theory defines things such as harm thresholds. What we as researchers collectively aspire to, in other words, is the support of something that *works* to make civil wars less likely and less destructive. So, for example, although in theoretical terms, we recognize that many of these measures are literally arbitrary, we can say that when the theories built on them help guide policy-makers to more effective policies, they remain a useful standard until or unless replaced by something that works better.

What *is not* a civil war?

Protests, riots, and mass demonstrations

Despite the lack of consensus on a definition of civil war, most academics and policy-makers can agree that civil wars demand organized, intentional violence, and this is what sets them apart from many other forms of lethal civil unrest. While they may also involve large numbers of civilians taking action against the government, protests and demonstrations are not generally considered civil war. But there is an important problem here. Once we accept that the state, as one form of political association, is chiefly responsible, at a minimum, for our physical security, we can understand a built-in bias against even non-violent protests and demonstrations. First, historically many demonstrations that start as non-violent can tip into looting (a form of violence against property,

33

not people) or arson and riots (which can easily cause death or serious injury). Second, what the "body count" measure highlights is the degree to which so many of us take for granted that lethal physical violence matters most; when in reality, leaders who justify their rule on a religious authority (such as Iran), or dynastic legitimacy (such as North Korea's Kims, or Syria's Assads) may react to criticism—ideas only—as if these were tantamount to acts of violence. The crimes of "blasphemy" or "heresy" are classic examples.

Gene Sharp, who wrote the seminal text on civil resistance, defines non-violent resistance as "those methods of protest, resistance, and intervention without physical violence in which the members of the group do, or refuse to do, certain things." As Erica Chenoweth and Maria Stepan catalogued, non-violent resistance can include boycotting certain areas or industries, staging sit-ins or extended protests at government offices, or coordinated marches. Sociologist Kurt Schock wrote that non-violent action "involves an active process of bringing political, economic, social, emotional, or moral pressure to bear in the wielding of power in contentious interactions between collective actors." The key concept implicit in both definitions is the idea of obedience and consent. Citizens voluntarily consent to government rule, but reserve the right to retract their consent.

Gandhi's leadership of the demonstrations that freed India from its British colonial rule and Martin Luther King Jr.'s promotion of non-violence during the American civil rights movement are two often-cited examples of successful civil resistance. In both, the protestors got their demands without resorting to violent methods, though they suffered violence at the hands of their adversaries. However, there are many other, lesser-known examples of non-violent civil resistance, including Timor Leste's independence from Indonesia in 1995 or the 2003 Rose Revolution in Georgia.

But non-violent action does not always work. Disorganization among leaders, a nuanced government response, or a decision by some protestors to engage in violence can reduce the effectiveness of civil resistance. In some cases, conflicts may begin as civil and non-violent movements, but quickly escalate into civil war. In the case of Syria, protests and peaceful demonstrations turned violent when there was a mass defection from the Syrian army with former government forces forming the rebel "Free Syrian Army" and solidifying the transition from civil resistance to armed insurgency. When that happens, it is time to start using battle deaths, organization, and level of violence to determine when exactly the civil war starts.

And in more recent times, rapid advances in surveillance technology have enabled governments to shut down non-violent resistance movements by identifying and arresting their leaders, or even just ordinary citizens coming to the streets to register their disapproval. In the Russian Federation since 2009 (and especially since March of 2022), Hong Kong in 2019, and Belarus in 2020 we have witnessed unprecedented success of governments using facial recognition backed by artificial intelligence to demobilize protests by identifying leaders and arresting protesters. This is new.

Genocides and mass killings

Similarly, although we often associate genocide with civil wars, genocide can occur in or out of a civil war. A genocide can happen at the beginning, during, or after a civil war, but is not the same as civil war. Genocides can happen in wars between states as well, but they generally share an important distinction: on genocides, there is no effective resistance by one side—government forces are committing the vast majority of the violence (sometimes through proxies, as happened when Slobodan Milosevic's Serb government delegated genocide to Serb paramilitary groups in Bosnia-Herzegovina, Croatia, and later Kosovo in the 1990s). While genocides are important features of civil wars in places like Rwanda or El Salvador, mass killings alone are not sufficient to meet the definitional criteria for civil war.

Terrorism

Terrorism is separate from civil wars. Often, just like with genocides and mass killings, the two overlap. Scholars use official definitions to clarify whether an event is a full-blown civil war or a terrorist insurgency, or if terrorism inflicts enough deaths on both sides or shows sufficient organization by the rebel group mounting the opposition to be considered a civil war. A few terrorist acts do not constitute a civil war. In fact, many consider terrorism to simply be a tool combatants use that is not unique or specific to intrastate wars. In *Resort to War: 1816–2007*, Meredith Sarkees and Frank Wayman wrote that "terrorism is not inherently a new type of war, but is a tactic that has been used in numerous wars throughout the 1816–2007 time period."

But here two additional key points emerge. First, as noted above, our collective tendency to give the state priority over non-state actors because we rely on the state to provide us with physical security tends to act as a bias that makes *any* act of resistance to the state appear to put the state's citizens at physical risk. Acts of violence directed against agents of the state—its military and police in particular—may seem to us a collective threat, even when these same agents of the state have been responsible for our oppression. It is for this reason that historically, we so often hear agents of the state defend acts of harsh repression as being in the service of "order" or "stability."

Second, although all agree that a core feature of terrorism as a tactic is that it deliberately targets civilians—including even children— whether we think an act of terrorism is criminal or political very often depends on whether we identify a targeted government as closed or open. When terrorists have legal means in an open political system of achieving the change they seek, we tend to think of them as criminals. By contrast, when challenging a closed political system—say, a brutal dictatorship that offers no non-violent alternative to resistance—we may think of the terrorists as freedom fighters. This last part is complicated by the problem that we have a hard time imagining that the use of an inherently illegitimate tactic could result in a just outcome.

Above all, terrorists aim to get *attention*. The hope is that the attention they gain from violence either directly mobilizes social support (thus shifting their insurgency from a tiny violent minority to a broad social resistance), or indirectly does so by provoking an incumbent government to overly harsh and indiscriminate reprisals.

How do civil wars relate to other types of conflict?

Continuing the discussion of what civil wars *are not*, how do civil wars differ from other types of wars, like interstate and extra-state wars? These other wars are not confined within the borders of one state, and as seen in Figure 3 on the next page, in recent decades civil wars have become far more common as compared with other types of war.

Interstate wars

Interstate wars are wars between two states, with statehood being defined—for example in the Correlates of War—as the possession of a territory, population, diplomatic recognition, independence, and sovereignty. The term "sovereignty" is shorthand for the idea that no one outside a state may legitimately tell another state's government what to do or not do. Interstate wars are the traditional type of conflict that comes to mind when one thinks of "war"—World Wars I and II are obvious examples. They also differ from civil wars in several ways. Combatants are usually from "regular" armies in interstate wars—they wear uniforms with distinctive insignia, and crucially, their leaders share definitions of victory and defeat. By contrast, the irregular forces most common in rebel groups—many without uniforms, who may fight at night and then return to "civilian" life during the day—typically play leading roles in civil wars. Stereotypically, interstate wars occur in the "developed world"—north of the Equator—while civil wars

Civil Wars

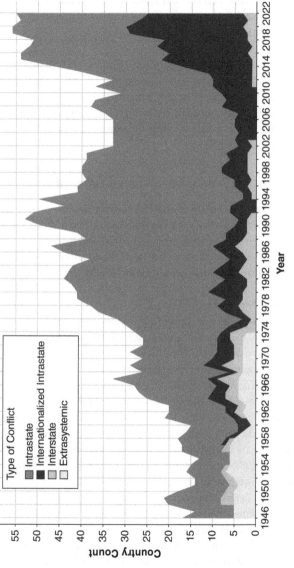

3. Armed conflict by type, 1946–2022

happen more frequently in the global South. And interstate wars since World War II tend to be less deadly and shorter than civil wars.

The frequency of interstate wars has dropped sharply since World War II. Two reasons are most often cited by scholars of international politics. First, the lessons of the years leading up to World War II were both that a collective security organization and organizations devoted to preventing economic disaster would be needed to prevent an even more destructive third world war. The United Nations remains a flawed institution, but together with such innovations as the World Bank, International Monetary Fund, and World Trade Organization, it has helped to prevent a third world war. Second, the Cold War fear of a global thermonuclear war between the USSR and the United States and its allies created a reasonable concern that even purely interstate wars might escalate, ultimately resulting in a species-ending nuclear war.

Tragically, the success in preventing a third world war did not mean worldwide peace. During the same period in which interstate war declined, the number of civil wars shot up. Since 1940, there have been about 136 civil wars, compared with 45 interstate wars. This difference in magnitude is seen in death totals too—since WWII about three million people have died in interstate wars, compared with 15 million in civil wars.

Extrasystemic wars

Extrasystemic wars involve fighting by a state outside its own borders against the army of an entity that is not considered a fellow state. Colonial wars and wars of independence are the most frequent examples of this phenomenon. Though they are distinct by definition from intrastate wars, civil wars can become extra-state wars when a state actor intervenes in another state's civil war. However, if that intervening state assumes most of the fighting against the local government forces, the war can then become an interstate war—as seen most notoriously in the Soviet interference in the Afghan civil war in 1979. As expected, this type of conflict occurred most frequently in the wave of decolonization following World War II but has since mostly tapered off.

Are modern civil wars different?

Even if the definition of "civil war" were settled, there would still be ongoing debates about the nature of civil wars over time and whether this is changing. Some scholars see a divide between post-Cold War civil wars ("new") and those before them ("old"), with a certain lack of legitimacy afforded to new civil wars. As Stathis Kalyvas summarizes in his article questioning the validity of the distinction, new civil wars are distinguished as a criminal phenomenon, while their predecessors, old civil wars, are considered political ones. The key differences between these two classifications come down to causes and motivation, sources of support, and level of violence. While such a divide may make it easier to study these new wars, as it implies that they require both new research strategies and new policy responses, the distinction is often a false one.

Causes and motivation

Old civil wars are often considered "ideological, political, collective, and even noble," according to Kalyvas. The causes of these conflicts were assumed to be collective, rational grievances, such as social change—often referred to as justice—and these grievances were well defined, clearly communicated, and universal. Cold War conflicts like the Salvadoran civil war, which pitted Communists against anti-Communist forces, are examples of such conflicts. In both, the struggle concerned ideology and whose worldview was expected to determine national governance. However, critics of this categorization charge that the importance of ideological considerations may have been greatly overstated in these conflicts because of the bias of the categorization's architects—academics who tend to be motivated by ideology themselves impart these ideas onto the participants and civilians they study.

By contrast, new civil wars are often viewed almost as criminal enterprises, with participants motivated by little more than private gain, greed, and loot. This greed is manifested in competition over control of state resources such as diamonds or petroleum. As former UN Secretary General Kofi Annan once pointed out, "The pursuit of diamonds, drugs, timber, concessions, and other valuable commodities drives a number of today's internal civil wars. In some countries, the capacity of the State to extract resources from society and to allocate patronage is *the* prize to be fought over." Conflicts in South Sudan over oil revenue, in Sierra Leone over diamonds, and in Liberia over rubber, timber, and other industries certainly fit this bill at first glance. The problem with this thinking is that economic concerns alone do not often cause civil wars, though they may motivate individuals to take up arms, and cash from their sale may sustain the fighting once a conflict starts. Moreover, as noted earlier, a lot of what motivates a person to risk life and family in civil war cannot be captured by easy-to-measure material things like bullets, exports, and so on. To actually know what drives participation and suffering we'd have to learn other languages, carefully interview combatants, and understand their culture and their memory of their own history. This cannot be done from afar solely with regression analyses on laptops in Chicago, Tokyo, or Oxford. Given this, the distinction between old and new is unlikely to be as useful as proponents argue.

Social scientist Mary Kaldor, however, has a different view of new civil wars. She writes that they are fought primarily in the name of identity (ethnic, religious, or tribal) and that the distinction between old and new wars lies, in part, in the role identity plays. In old civil wars, identity was an instrument of war, but in new civil wars, it is the aim of the conflict itself. Kaldor also emphasizes globalization and technology, rather than ideology.

Harkening back to the work of Karl Polanyi, who imagined a global South backlash against market capitalism, she calls new wars "wars of the era of globalization," and writes that they typically take place in areas where states have been greatly weakened because of opening up to the rest of the world.

Sources of support

Old civil wars seemingly garnered widespread popular support because their causes were noble and justified, and were therefore "obviously" shared by the rest of society. However, old wars can fall victim to the same elite-based definitions as new wars do—elites articulated the reasons for rebellion as national cleavages and observers simply assumed that there was popular support mobilized along those cleavages. In reality, this may not have been the case at all. When one dives deeper into the rationales of citizens, it is evident that popular support was, in fact, won and lost during the war itself, rather than being static and purely consensual. Further, Cold War dynamics often ensured that both sides of these ideological old wars would be well funded by the United States and the Soviet Union, with such superpower patronage bestowing an air of legitimacy on both sides in old wars that is not as evident in new ones.

New civil wars, on the other hand, appear to be fought by minority rebel groups who lack any such support and are acting only for their own gain. While this may be true in terms of some terrorist insurgencies or local rebellions, such as Boko Haram in Nigeria or Islamic State-affiliated groups in Indonesia, this certainly cannot be said about all post-Cold War civil wars. The Mozambican Civil War (1977–1992, 2013–2019) began with an insurgency (RENAMO) supported by then-Rhodesia (later South Africa). The first war was "old" in the sense of enjoying outside support and being fought along ideological lines: RENAMO rejected the government's (FRELIMO's) socialist policies and fought for the reinstatement of tribal leadership. But the second war, though post-Cold War, was hardly about lootable resources or greed. It engaged the same issues as the 1977–1992 civil war. And like too many civil wars, "old" or "new," the Mozambican Civil Wars also took a terrible toll on civilians.

Levels of violence

Lastly, the (perhaps false) distinction between new and old wars is that old ones were more "civil," if you will, with violence better controlled and limited to the concrete end of "winning" the war. Examples of old civil wars that upheld this notion are the American, Russian, and Spanish civil wars, which are praised for the presence of regular armies, command structures, and clearly articulated strategies on both sides. In all, the violence was seemingly organized, with the state retaining some sense of control. The violence of new wars, on the other hand, is characterized as horrific and senseless, and carried out with no agenda whatsoever. Grotesque accounts of massacres in Sierra Leone, Algeria, and elsewhere made violence seem completely random and the ultimate aim of the perpetrators being simply to create chaos rather than achieve any meaningful political change. But after a closer look, many of these assumptions do not hold. The violence in Algeria, for example, was found to be highly planned and selective.

We should consider two additional risks in any attempt to categorize civil wars as "old" or "new." First, it may be that as microprocessors became more inexpensive in the 1990s, increased demand for machine-readable data caused a shift in focus from difficult-to-digitize human motivations and aspirations (subjects), to material objects such as cash, corpses, gold, and diamonds. Second, we need to acknowledge the risk that civil war scholars from the global North may be inadvertently underlining a self-congratulatory ethnocentric judgment that civil wars in the global North were "civilized" and about ideals, whereas those in the global South are "savage" and about greed.

Conclusion

Many of the differences between so-called old and new civil wars do not actually stand up to much scrutiny. Old wars are not as pure, just, popular, and civil as they are often remembered. And new wars are not as illegitimate, unwarranted, wanton, and senseless as some believe. Both old and new civil wars share many similar features, with groups and their leaders often acting rationally and strategically to try to defeat their opponents and achieve the objective of asserting power over their "own" territory while many suffer the consequences. A civil war is still a war within a state regardless of when it occurred.

But this does not mean that differences do not exist. Global politics, technology, and international norms all affect how wars play out. So, as these things change, wars and their consequences can change too. There are long-term trends in the most common causes of civil wars and we can see differences in what is most frequently fought over in different eras. Some causes are shared among most civil wars, while other features are more commonly associated with specific types of civil wars.

Chapter 3
Causes of civil war

With the definition of civil war so intensely debated among academics, analysts, and policy-makers, it should come as no surprise that the reasons why these conflicts occur are highly disputed as well. There are even disagreements over how to classify different types of civil wars. Still, this debate is relatively new in civil war scholarship. A focus on the different types and risk factors of civil war only emerged in the early twenty-first century. Up until then, studies mainly focused on risk factors without considering that there may be more than one "type" of civil war.

In 2001 Nicholas Sambanis was the first contemporary scholar to systematically analyze differences between two unique *types* of civil wars—ethnic and nonethnic—and the very different factors that commonly trigger each type. In "Do Ethnic and Nonethnic Civil Wars Have the Same Causes? A Theoretical and Empirical Inquiry," Sambanis found that there were significant differences in what caused "identity" and "non-identity" civil wars.

But before we discuss the different causes of "identity" or "ethnic" civil wars versus "non-identity" or "nonethnic" wars, it might be helpful to delineate some factors that are common to the onset of all types of civil wars. Five are worth introducing: a history of prior conflict, the role of elites, geography, demography, and governance.

The first general cause all types of civil wars tend to share is that they come in cycles. In other words, if a country suffered a civil war in the past, then it is more prone to civil war in the future. Even if a past war seemed to settle previous disputes, old animosities have a way of returning. Moreover, the more recently the last civil war ended, the more likely war will recur. The Mozambican Civil War (1977–1992, 2013–2019) is a case in point. Therefore, avoiding civil war altogether is critical to stop future civil wars, and countries are most at risk shortly after a previous civil war.

Elites are a necessary condition for civil wars to emerge, but they are not sufficient—there are always other factors that cause a war. Elites are a necessary because they help coalesce identities and grievances. In states with fragile institutions and electoral practices, elites often ramp up ideological discontent over class issues (including things like land ownership), harden the distinctions between different identity groups, or strengthen identity ties in the form of nationalism or more extreme hyper-nationalism.

Citizens within a society are soon forced to take sides even if they would prefer to sit on the fence. Oligarchic control of land protected by a history of repressive military rule against an impoverished and dislocated peasantry led to war in El Salvador, while the wars in the former Yugoslavia emerged as contests among elites eager to dominate Yugoslavia in full (Serbs) or as separate entities (Croats and Slovenes), leading to the division and eventual destruction of the country along nation-state lines.

Geography and demography also matter. There are simply some 'bad neighborhoods', to quote Myron Weiner. States that border states with a civil war are more likely to experience civil wars themselves. In terms of demographics, countries with so-called youth bulges (a higher-than-normal number of young people), large populations, or gender inequalities are more prone to war. Societies with a disproportionate number of males in the age range of roughly 15–29, indicating a youth bulge, have a higher propensity for civil war. After all, frustrated unmarried young males are common recruits for warring factions. The size of the country and its total population also matter, with larger populations heightening the risk of civil war. This makes sense since people are a form of power that can be mobilized to challenge an existing political order. And finally, sexual and gender equality make a difference. Societies that afford females the same opportunities as males are less prone to civil war. Respect for basic rights is thus critical.

Relatedly, governance is the last general factor to consider, with the effectiveness and type of government affecting the chance of civil war. Countries with weak governments, weak rule of law, and high levels of corruption, typically thought of as dictatorships face a higher risk for civil war. The likelihood of a civil war thus depends to some extent on whether the government is a democracy, autocracy, or anocracy (a type of regime that falls between democracy and autocracy).

Consider a graphical representation of this relationship in Figure 4. On the left axis, the likelihood of civil war erupting moves from low to high, while on the lower axis is the regime type, moving from autocracy on the left, to anocracy in the middle and democracy on the right. What we find is a parabolic relationship: democracies and autocracies have a lower probability of experiencing civil war, while anocracies have a high risk. Why?

Democracies include checks and balances on the executive and are generally representative of the people, who have a say in their government. The leadership in autocracies is fairly free to run the government without encumbrance from the population. But anocracies fall between these types, resulting in anocracies as the most prone to civil war.

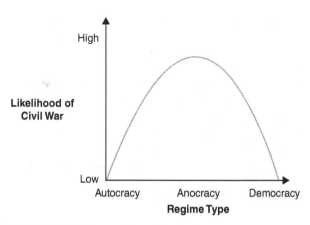

High

**Likelihood of
Civil War**

Low

Autocracy   Anocracy   Democracy

Regime Type

4. Probability of civil war across regime types

While these five factors are common at the onset of all civil wars,
there are also distinct factors that help explain the emergence
either ethnic or nonethnic wars.

Ethnic civil wars

Consider your own personal identity—do you hold allegiance to
the state where you are a citizen or where you live? Or to a group
whose common history, language, culture, or values you share? Or
to a religious organization whose doctrine you believe in? These
questions are by no means exhaustive and are not mutually
exclusive—a person can simultaneously feel a part of one state
while also identifying with a specific religious group within that
state, or even beyond the state's boundaries. But these questions
help us reflect on what identity is, how it is defined, and then how
identity becomes entangled in the onset of civil wars.

There are two basic debates in the literature about the origins of
identity, including ethnic identity. One camp holds that identity is
fixed, and that attachments and sentiments are the fundamental,

unchanging foundation upon which group economic, social and political interests and claims rely.

The competing view believes identity as malleable, seeing it as the product of a particular time, place, and series of events. In practice, however, these approaches often combine into a sort of a hybrid that considers ethnic affinity to be universally latent that manifests actively and politically when a group identifies a common threat, one that is perceived or actually is a threat to the group's existence, integrity or interests.

But even if these views are reconciled, the question still remains— what is an ethnic group, exactly? According to Anthony Smith, a leading scholar of ethnicity, an ethnic group is a group of individuals who share (1) a common trait such as language, race, or religion, (2) a belief in a common descent and destiny, and (3) an association with a given piece of territory. An example of this is seen in the competing definitions of "Serbs" and "Croats" during the Yugoslav Wars in the 1990s. "Serbs" saw their ethnic brethren as sharing a common language of Serb, common religion of Orthodoxy, and a common homeland in and around contemporary Serbia, Kosovo, and parts of Croatia and Bosnia. "Croats" shared the common language of Croatian, the Catholic faith, and a homeland of Croatia.

So how does an ethnic group reconcile itself with the concept of nationality, if at all? Clearly, in the Yugoslav Wars, Serbs and Croats did not agree on a shared nationality, believing their ties to their respective ethnic groups were stronger than those to an overarching state. However, this is not always the case, and these two ideas often co-exist. In the United States, for example, "hyphenated Americans," an epithet first used to disparage foreign-born Americans, is now common parlance. An individual can refer to themselves as "African-American," "Irish-American," or "Indian-American" as a way of expressing their heritage, without their nationality and loyalty to the American state being questioned. According to Walker Connor, nationality is different from ethnic groups in that it entails self-recognition—each member recognizes their own membership in the nation, the membership of fellow co-nationals, and the "non-membership" of non-nationals, and the nation shares a common objective that identifies it as distinct from other groups. An ethnic group can be considered a latent nation, perhaps not quite activated or mobilized, whereas a nation is a politically active ethnic group that demands greater cultural autonomy or self-determination.

What is "ethnic war," then? At its root level, ethnic war means groups of people fighting with other groups, where the "other" is usually defined in terms of race, language, or religion, resulting in large-scale, organized violence. There are several patterns that ethnic wars follow. In some cases, one of the groups may dominate the state and use state institutions and resources to defeat a competing ethnic group, as was witnessed in the Rwandan genocide. Or the war might involve two ethnic groups fighting each other without the involvement of a central state. Lastly, while such a war may or may not involve territory, it often does. In fact, from 1940 to 2000, 98 percent of civil wars fought for territorial control were incited by ethnically based demands, while nearly two-thirds of all ethnically based civil wars involved fights over territory.

A brief history of scholarship on ethnic civil wars

To appreciate the unique causes of ethnic civil wars, it is worth discussing how the study of civil wars has changed over time. Not surprisingly, there was little focus on intrastate wars prior to the 1960s, as the century had largely been defined by massive, interstate wars like World War I and II, and concern over the possibility of thermonuclear war between the Soviet Union and the United States.

It was not until after the fall of the Soviet Union and end of the Cold War that people began to accept the idea that identities mattered—largely because the breakup of the USSR had come from within; and because the theories that had been applied to wars between states did not seem to help much in explaining or predicting wars within states. Groups within the state acted in a way that fundamentally altered the reality of interstate politics. This reality, combined with the growing and proximate horrors of the ethnically fueled Yugoslav Wars and the technological revolution lowering the costs of mass migration from the developing to the developed world, encouraged more scholarship of ethnic and sub-state violence and cemented ethnic civil wars as a proper research subject in its own right.

With the end of the Cold War, research on ethnic substate violence really picked up. For the first time, ethnic groups were conceived of as rational actors who responded to fear and opportunity much like traditional states did—though this was far from the whole story. At the same time, economic-driven arguments for ethnic conflict were becoming more mainstream. In my own book, *The Geography of Ethnic Violence*, I explored why, given similar existing conditions, ethnic civil wars broke out at certain times and places and not others. My research showed that when ethnic groups were concentrated in what they identified as their homeland, they were dramatically more likely to demand increased autonomy when a crisis weakened a multinational state's government. The collapse of the Soviet Union in 1991 proved to be just such a crisis. For Tajikistan, a majority of whose ethnic Tajik population remained in the diaspora, that demand meant better trade terms with Russia; whereas for Chechnya, in which over 90 percent of ethnic Chechens worldwide lived, the demand was for independence from Russia.

Causes of ethnic conflict

There are several prominent theories that seek to explain why ethnic civil wars emerge—ethnicity and nationalism, economic grievances and greed, manipulation by elites, religion, and, rationalist explanations like the security dilemma and commitment problems. Often, these work together to cause an ethnic, intrastate conflict.

Ethnicity and nationalism

It seems obvious that ethnic civil wars would be caused in some way by ethnicity and nationalism. In ethnic civil wars, ethnicity almost acts as a gateway to interstate conflict—or a war between two states—as it fractures society along ethnic lines, further exacerbating the effects of fear, greed, or elite manipulation. However, it can also be a cause in and of itself, as was seen in Rwanda—one of the most infamous ethnic conflicts of the last century—where an estimated 500,000–800,000 people were murdered just because of their ethnic group affiliation.

Although most scholars accept that politics and elite maneuvering are crucial to the dampening or inflaming of the salience of ethnic ties and that most ethnic groups manage to co-exist most of the time, the basis of ethnic ties themselves is debated. Today most scholars exploring the relationship between ethnic identity and civil war argue that ethnicity is a social construct that is often created or amplified by elites or private interests in the service of a particular agenda. Stuart Kaufmann, in his *Modern Hatred* book, for example, speaks about the importance of group prejudice, and the ways in which hatred is inflamed through politics and the manipulation of an ethnic group's myths and symbols, evoking emotional responses amongst the population. Regardless of whether these ties are essential to an individual who is born into an ethnic group and/or manipulated by elites for political or economic purposes, ethnic differences can lead to war.

When an ethnic group, particularly one that is seeking to uphold its established status and authority, faces challenges due to differing demographic growth rates among other groups, shifts in the equilibrium can emerge, potentially leading to conflicts and wars. In many places, larger groups have control over scarce resources, such as jobs or offices, so when a group used to control finds itself in decline relative to other groups, it may resort to repression or violence to protect its privileges. It also follows that the group or groups that are gaining may begin asking for a greater share of privileges in proportion to their gains.

Security dilemma

The security dilemma, a well-known international relations concept, has been usefully applied to the study of ethnic civil wars and their causes. In the security dilemma—a situation in which one group cannot increase its own security without making another group fearful of injury—the central driving force is fear. Think of this dilemma in terms of characters in a typical elementary school playground—the big kid, the little kid, and the teacher. When the teacher is around, the little kid has no fear of the big kid (who in any case might be a nice person). But should the teacher step inside, the vulnerable little kid, not certain that the big kid is in fact a nice person, will worry that if the big kid becomes aggressive, the little kid might get hurt.

Similarly, in a multiethnic state, when the central government loses power, it can no longer protect the interests of ethnic groups, and these groups begin to fear for their safety. The decline of the central state creates a vacuum in which ethnic groups compete to establish and control a new regime that will protect their interests. Considering the future composition of a new government that is dominated by opposing groups, and the probable treatment of their own group within such a new regime, ethnic groups fear widespread discrimination and even death. Adopting a worst-case scenario logic, each group assumes other groups have offensive capabilities and hostile intentions toward them. As each group mobilizes and arms in self-defense, the probability of war increases. The destruction of Saddam Hussein's control of Iraq is illustrative here. For Sunnis in Iraq, who made up a significant demographic minority but who controlled everything of value in the Iraqi state, the prospect of democracy after the fall of Hussein was terrifying. Having oppressed the majority Shi'a population for decades, Sunnis suddenly had to fear what a Shi'a governed Iraq might bring to them. It was this more than any other factor that led Sunni groups to attack US and Coalition forces in Iraq, as well as Shi'a elites.

Elite-driven

In ethnically divided societies, or really in all societies for that matter, power is often concentrated at the top with the elites who hold significant decision-making power over the common citizen and disproportionately reap benefits of governance. Elite-driven explanations of ethnic civil war emerged after seeing the impact of charismatic strongmen such as Franjo Tudjman and Slobodan Milosevic in the 1990 Balkan Wars and the political elites in several countries in Africa, where many conflicts include ethnic divides that are manipulated from above. Some elite-driven explanations for ethnic civil war focus on the material incentives like oil rents, while others draw on nonmaterial appeals, such as a particular use or retelling of history in the pursuit of national or personal aims.

Though it has its flaws, the popularity of the elite-driven approach is easy to comprehend. First, it flatters a deeply held conviction that people, like children, are generally good, and therefore bad behavior is best explained by bad leaders, teachers, or parents. The blame lies with the elites, not with the rest. Second, it is simple, often too simple. What caused World War II? Hitler. What caused war in Yugoslavia? Milosevic. The main weakness of the elite-driven approach is that it relies on a necessary gap between national aspirations and the will of the leader in that it denies the possibility of representative leadership, or that the leader is acting on behalf of an ethnic group and voicing widespread beliefs. Additionally, this approach cannot explain why leaders who possess similar amounts of charisma and are just as good at public messaging experience quite varied levels of popular support— some fail, and some succeed beyond even their highest hopes.

Grievance and greed

Think again of the dynamics of the elementary school playground. While often the big kid isn't a bully, when he is he may pick on little kids for their lunch money. Though more complex than this analogy suggests, the causes of ethnic conflicts are occasionally similar. Ethnic groups may resent other groups for their material possessions and resort to violence to level the playing field.

Early theories of the impact of greed and grievance on ethnic conflict focused on relative economic development and the resource distribution between ethnic groups. In both cases, scholars predicted that violence emerges when one ethnic group blames another for a decline or lack of access to economic resources in their respective territories or through government distribution. Disparities in economic development lead to the belief that ethnic groups are not able develop their full potential and that other ethnic groups are somehow responsible. Resentment and fear builds, which can then result in violence between the two groups.

Beginning in the 1990s, scholarship on the impact of grievances expanded, but in a way that largely took ethnicity out of the equation (more on that later). But more recent scholarship returned to the impact of economic grievances on ethnic conflicts. In 2013 Lars-Erik Cederman, Kristian Skrede Gleditsch, and Halvard Buhaug found that relatively poorer groups in a country fight more wars. They disputed previous studies that questioned the importance of grievances about political or ethnic inequalities, pointing to the experiences in South Sudan, Myanmar, and Yugoslavia. The complicating factor in their research is that in many but not all places in the world, political power *means* economic wealth. Nevertheless, it can be difficult to tell whether a group pursuing more political power is seeking power for its own sake, or seeking power as a means to make up for years of forced poverty and government corruption.

Political systems and governance

It makes sense that certain government structures—usually representative ones like federal or parliamentary systems—are theoretically better suited for governing ethnically diverse populations. They are inclusive and attempt to reach consensus. They are thus better at resolving issues without violence. Furthermore, when groups lack necessary resources, these systems provide a sound structural solution to make distribution of economic resources and political offices more equitable. This is a popular idea in the West, as it seems to confirm the virtues of democracy and the validity of the "democratic peace theory," which postulated that liberal democracies do not go to war with one another, in part because they are accustomed in their interior politics to resolving disputes short of violence. The connection of a political institutions approach with ethnic war is clear: where mature democratic institutions are in place, we can expect to see not only a reduction in interstate violence, but also a reduced likelihood of intrastate violence.

Ethnic tensions can escalate into violence when political systems are under attack or have failed completely. The "failure" of Yugoslavia's federal system to contain the war in the 1990s, or the Lebanese national government's inability to ease sectarian tensions in 1975, provides evidence that a breakdown occurs when ethnic groups are disenfranchised within a state system. Therefore, authoritarian one-party regimes are susceptible to ethnic conflict given that they deny opportunities to some ethnic groups to participate in governmental decision-making processes. Yet, as shown earlier it is not perfectly authoritarian or democratic regimes—anocracies—that are the most war-prone. While autocrats seem more willing to use brutality—President Bashir Assad in Syria in 2011 is a good example—democratic systems have institutions in place to allow for voicing of grievances and challenges to the existing political order.

Religion

The impact of religion on ethnic wars is disputed but should not be ignored. In recent years, the role of religion in intrastate conflict has been increasing—one can look to conflicts between Hindus and Muslims in India, secularists, communists and the Taliban in Afghanistan, or the Islamic State's role in the Syrian Civil War as prime examples. Some scholars argue that because religious conflicts concern indivisible goods, religious groups are less able to negotiate or compromise and thus more likely to resort to violence. Others believe that religious violence stems from disappointment with the secular state. Religion inverts the assumption that ethnic groups in civil war are rational actors, as the promises of martyrdom undermine two pillars of the state system—bargaining and deterrence.

My own theory of the role of religion builds on Jack Snyder's model of nationalist outbidding, or the idea that political elites will attempt to outbid each other to appear more nationalist than the other and thus win the approval of the people. The individual who wins the contest will be established as the most credible defender of the nation and will receive the resources required to maintain their leadership. Religious outbidding is a similar phenomenon—elites outbid one another to establish their religious credentials and establish support to counter an immediate threat. A critical difference today, when so many nations are entirely bounded within states, is that as religious elites they have transnational reach, which means that they can appeal to both domestic and external audiences for support, as Muslim fighters have demonstrated with their solidarity and engagement in Afghanistan, Bosnia, Chechnya, Iraq, and Syria, to name a few places.

As with the other possible causes of ethnic civil wars, religion is often one of several factors at play. Nevertheless a civil war is more likely to become a religious civil war when four conditions hold:

(1) the government or rebel leaders are immediately threatened; (2) the resources (such as small arms, cash, skilled fighters, and logistical support) needed to reduce or eliminate the threat may be acquired by framing a conflict in religious terms; (3) the society has pre-existing, though not necessarily deep, religious cleavages; and (4) the government controls public access to information. In the Second Sudanese Civil War, for example, the Muslim central government attempted to impose Shari'a law on Sudan's non-Muslim citizens. This led to violence and eventually a destructive civil war that split the state in two, creating a new state: South Sudan.

Nonethnic civil war

Civil wars of this type also share a pattern. By definition combatants are likely to identify with an ethnic or nationalist group, but in nonethnic civil wars what motivates them to fight are things like ideology and control of natural resources. In other words, their ethnicity or the ethnicity of their opponents isn't central.

Economic concerns

Some scholarship on civil wars in the 1990s focused on economic interests and prosperity, basically taking ethnicity out of the picture. The primary rationale of these scholars was that rich states are less likely to experience civil wars than poor ones. War is considered an expected outcome of a utility calculation in which rebel groups are rational actors—potential rebels evaluate the potential gains to be had against the rewards of working a regular job. In many if not most places in the world, given the lack of economic opportunity, the risk of taking up arms to redistribute wealth from wealthy and corrupt government officials is worth it.

Conducting research for the World Bank, economists Paul Collier and Anke Hoeffler connected poverty and war, claiming that the origins of civil war are related to the share of primary products in the country's gross domestic product. Their work stressed the importance of opportunity costs—poor states have more unemployed people with less to lose who are more likely to take up arms against their government. This is the same logic that mobilizes the rebels in Suzanne Collins's fictional District 12 against a corrupt and repressive Capital in her *The Hunger Games* novels. When a status quo leaves little to lose, the risks of armed insurgency against incumbents drops to almost nothing.

In a similar vein, James Fearon and David Laitin identified per capita income as the best indicator of civil war, finding a negative correlation between wealth and war. According to their research, income levels were better at predicting civil war than other factors including political grievance and ethnic diversity. These findings, however, have since been walked back to conclude that economic incentives remain but one of many factors causing civil wars. For example, Barbara Walter, also using statistical analyses and focusing on civil war in physical or kinetic terms, demonstrates that the conditions present in contemporary US domestic politics are strongly associated with the onset of civil war, thereby demonstrating that even in an advanced industrial and technical economy with a written constitution and checks and balances against authoritarian rule, civil war is possible.

A main feature and problem with a number of these studies is that they limited their analyses of civil war to a single indicator, like GDP. This is a problem because aggregate numbers do not account for sub-state or regional level differences, and this is precisely where intrastate wars often begin. One region of a state might seek to become independent because it is poorer and feels as if it is not getting its fair share of the state's coffers. Yet the measure of GDP would not capture regional differences in economic development, poverty, and wealth. Further several studies do not distinguish between ethnic and nonethnic civil wars, and thus find no meaningful difference between the two types of wars.

Resource imbalance

Like the economic explanations, the idea that resources such as oil or minerals can cause civil war became quite popular, though it has been similarly scrutinized. Conflicts in Africa, in places like Sudan, Sierra Leone, and South Sudan, over commodities such as oil, gems, minerals, and timber seemed to confirm the hypothesis that resources and wars were closely linked. Indeed, the UN Environment Program estimated in 2009 that at least 40 percent of all intrastate conflicts in the past 60 years were linked to natural resources. That greed and the unequal distribution of resources drive conflict is appealing because of its simplicity—there is a clear economic rationale for war and thus a relatively easy solution compared with all of those of messy ethnic conflicts.

But the simple explanation that resources cause conflict is disputed. In fact, resources have been shown to be only an enabler, rather than a cause. In *War and Conflict in Africa*, Paul D. Williams writes that civil war scholarship tends to focus too much on the rebel side of the equation, when most violence is initiated by state elites. Additionally, resources are required for waging war—getting the people, money, and weapons necessary to fuel an intrastate rebellion—and thus it makes little analytical sense to talk about resource wars, because non-resource wars are not possible.

Who fights?

Individual motivations for joining a war are also important, and of interest to scholars. Beyond determining why ethnic and nonethnic civil wars break out, there is a need to discover why civilians choose to get involved. A wave of scholarship emerged in the past decade examining this very question—why do individuals, who could choose to stay neutral in an intrastate conflict and enjoy greater safety, decide to run the considerable risk of supporting insurgents? Like much else in this field, the answer is highly debated.

Many explanations rely on Mancur Olson's seminal *The Logic of Collective Action*, which proposed that people are unlikely to join groups in pursuit of a shared benefit absent selective incentives or certain conditions—otherwise they will "free ride" on the actions of others. But what exactly are these conditions? Some researchers believe that material grievances, such as the lack of access to education or employment, in addition to potentially causing civil wars, can also motivate individuals to support violent groups.

Other hypotheses that have been explored include the degree of control each group asserts over a territory, an individual's pre-war level of political involvement, or the desire to right existing wrongs and seek revenge. As a long-suffering peasant-turned-insurgent of El Salvador's brutal civil war stated, "This is what I think: what was the war for? For the solution to the land problem. We feel something already, and we're sure that we will be free—that is a point of the war that we have won. Higher incomes? Who knows? But that we not be seen as slaves, that we've won."

Conclusion

While disagreements remain among scholars over competing explanations for what causes civil war, we know a lot more today about the common factors that most states share before war breaks out. Countries are at a greater risk if they have a history of conflict; feature political elites who try to exploit and harden divisions between different identity groups; are adjacent to other states experiencing war; possess large populations with a youth bulge or gender inequalities; and critically, are led by a government that is weak, corrupt, and anocratic. But we also know that there are differences between ethnic and nonethnic civil wars. Issues of identity drive ethnic wars, while nonethnic wars are more likely motivated by material concerns.

Regardless of why a civil war starts, it will inevitably cause death and destruction; and they seem even more awful because they are so much harder on civilians than interstate war. This is true for both ethnic and nonethnic civil wars. The immediate and long-term consequences for people and the states in which they reside can be immense.

Chapter 4
Consequences

Civil wars are devastating. Millions of people have died in civil wars since the end of the Cold War, over 10 times more than the number killed in interstate wars. Most refugees today flee civil war violence, with millions more forcibly displaced from their homes inside their own countries. Civil wars also complicate efforts to prevent disease outbreaks and collapsing healthcare systems cost additional lives. People suffer atrocities such as mass rape, torture, and mutilation. Jobs and livelihoods are lost. Economies falter and famines emerge. Schools close and infrastructure crumbles. Insecurity spreads beyond borders. And political and civil rights tend to suffer as governance worsens and regimes crack down on dissent and society writ large.

While the effects vary from civil war to civil war, the damage harms countries and the people living in them. Hundreds of thousands of civilians and combatants were directly killed during battle in the civil wars in Afghanistan, Cambodia, and Ethiopia. When including the numbers who died from things like disease and starvation because of the civil war in the Democratic Republic of Congo, estimates go higher than 5 million, making it one of the deadliest crises since World War II. More than half of Syria's pre-war population was displaced during its civil war, and millions ran from their homes during civil conflicts in Colombia, South Sudan, and Yemen.

But these are just the immediate consequences of civil wars, as there are also long-lasting harmful effects that can plague countries and people for years after the fighting officially ends. Old rivalries and longstanding problems do not magically end when one side wins or a temporary truce or compromise peace agreement is reached. New governments face the seemingly impossible task of trying to build a new political system, recreate a broken economy, and provide for a destitute and wary population in an insecure environment with meager resources at hand. Economic growth rates can often return to pre-war levels shortly after wars conclude, but war-damaged infrastructure including roads, schools, and healthcare facilities must be rebuilt, which takes years. Lingering hostile feelings can poison or deadlock the political system, often causing civil wars to recur. Previous civil wars increase the likelihood that a country will wage civil war again. Fifty-year-old adults still living in places like Azerbaijan, Chad, Iran, Myanmar, and Sri Lanka have survived five or more instances of civil war.

Clearly civil wars have both immediate and long-term effects on citizens' lives and livelihoods and a country's economy, governing structures, and stability. In this chapter, we will discuss the consequences of civil wars. We will first examine the impact of civil war on the civilian population in the short term, before exploring the long-term implications for the country's political and economic landscape. A central part of these long-term outcomes are the proximate causes of civil war recurrence after violence subsides, which include policies, such as partition, that states adopt with mixed success to prevent future war. Implicit throughout this discussion are the various attempts by outside actors, especially the UN, to limit the severity or duration of war and ensure the stability of peace once the war is over. The consequences of civil wars are both severe and long-lasting, and the international community would be wise to recognize the enduring legacies of civil wars when determining how to respond before, during, and after warfighting.

Death and displacement

The devastation and destruction caused by civil war is felt most acutely by civilians. The World Bank estimates that over 90 percent of casualties from armed intrastate conflict are civilians, with higher-than-average mortality rates continuing even after the conflict ends. Civilians are also forced to flee their homes in large numbers to seek safety in other parts of the country, or even outside of the country. Civilians can be directly targeted in wars depending on the objectives and tactics of the warring parties, and they can also be collateral damage—that is, killed unintentionally.

The numbers of people who are directly killed and displaced in wars depends on both the intensity and severity of the violence and the duration of the conflict. Particularly intense civil wars, like those in Vietnam, Korea, and China, can kill millions of people in a hurry, while the numbers of deaths and displaced persons will slowly add up in the wars that stretch on for decades without any side winning or a peace deal reached, like the ones in Colombia, Afghanistan, or Myanmar. Figures 5 and 6 show the deadliest and longest civil wars.

Some attempts have been made to illuminate why certain conflicts result in more battlefield deaths than others. Laia Balcells and Stathis Kalyvas, for instance, propose that when the incumbent government can marshal disproportionate resources against a smaller and lightly armed guerilla force there will be more civilian and combatant deaths. When the two sides are more evenly matched, wars between them are far shorter and less deadly. In terms of civilian deaths, Jeremy Weinstein argues that armed groups with access to income from natural resources are more likely to target and harm civilian populations, with deadly violence spiraling out of control when groups like this are active in places like Mozambique and Peru. The logic here is that normally, insurgent groups depend on social support for the resources they

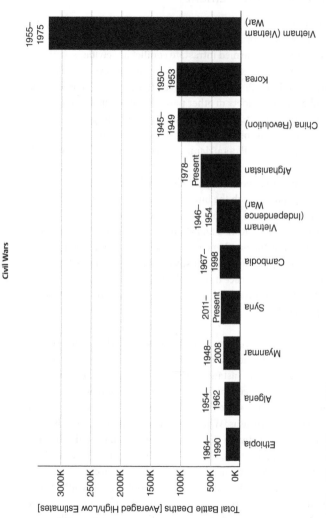

Civil Wars

5. **Deadliest civil wars: 1945–present**

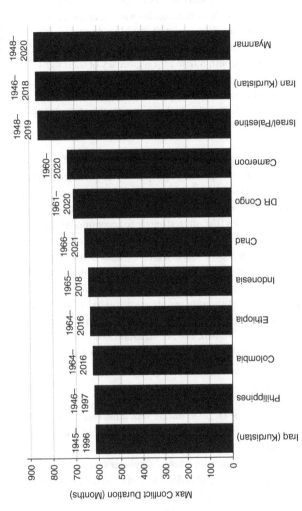

6. Longest civil wars: 1945–present

Categorized by UCDP/PRIO Armed Conflict Dataset Conflict ID

need to survive and make gains against incumbents. But with access to gold, diamonds, and oil, social support becomes less important. This is why Mao Zedong also warned against insurgent access to external material support: it would lead to the dangerous situation in which insurgents abused ordinary civilians (up to and including genocide), both hurting the legitimacy of their fight to defeat the government and making them vulnerable to destruction should the outside support stop. And as shown in Figure 5, the most destructive civil wars listed all included outside support.

Civilians are also forced to leave their homes in search of safety, and in recent years the number of people displaced has continued to rise. Tens of millions of refugees have fled their home countries, seeking shelter abroad. The majority of refugees migrate to neighboring countries that often struggle to accommodate the rapid influx of people. But even larger numbers of people are displaced within their own country, altering local demographic balances among different identity groups. These population shifts can affect an ongoing war and complicate possible solutions for ending the conflict. Millions of people are displaced within their own countries because of civil wars in Syria, Democratic Republic of Congo, Colombia, Yemen, and Afghanistan. Adults need to find new jobs, while youth can miss out on years of schooling.

Health and disease

While civilians can be targeted during war, most non-combatant deaths are not directly connected to violence. Instead, increased exposure to disease, lack of medical care, food insecurity, destruction of infrastructure, internal displacement, reduction in government spending on public health, and overall deterioration in socioeconomic conditions play a major role in mortality. In Darfur, the World Bank estimated that 90 percent of excess deaths were not a direct result of violence. Infant mortality also spikes during a civil war and stays elevated up to five years after a conflict ends. For the survivors of civil war, psychological trauma results in depression and PTSD symptoms for many years to come.

Employment and education

People are often compelled to change jobs or find new sources of income during civil wars. The effects vary. But the tendency is for formal jobs to fall as official businesses are destroyed while informal jobs, both legal and illegal, tend to rise, and the long-term earnings of those affected by war can suffer. Investment all but stops, because no one can be certain of what will happen the next hour, week, or month.

Internal migrants often find new informal gigs, although often for very low pay, after relocating and sometimes turn to illicit sectors including smuggling, drug production, and arms trading. Afghan farmers often relied on poppy cultivation and opium exports to make money during war, whereas Colombian farmers planted coca, fueling cocaine exports.

School-age children are particularly vulnerable during civil wars. Schools and teachers are often targeted by warring factions, and civil wars tend to close schools, decrease the average number of school years for local children, and reduce literacy and other education assessment measures, harming the future earning potential of younger populations. Girls seem to be the worst affected in many war.

According to UNICEF, millions of children are out of school and hundreds of educational facilities and personnel have been attacked during the civil wars in Syria and Yemen. More than two-thirds of primary and secondary teachers were killed or displaced during Rwanda's civil war, Cambodia was left with almost no trained or experienced teachers after its war, and most secondary teachers returned to Indonesia during the war in Timor Leste, leaving the soon-to-be-independent country behind. After

conflict, 45 percent of schools in Mozambique, 50 in Bosnia and
Herzegovina, and 85 percent in Iraq required repair or
reconstruction. Civil wars therefore affect both immediate and
long-term education levels in the country.

Economy

Civil wars hurt a country's economy, often causing production to
shrink, and leaving the country years behind more peaceful states.
GDP per capita tends to be significantly lower after the conflict
than before, with allocation of government spending toward the
military a significant drain on the economy. Drops in income per
capita are also caused by the destruction of physical and human
capital, diversion of foreign direct investment, reduced domestic
investment, loss of homes, and a breakdown of public health as
discussed earlier. There is reduced domestic investment with
long-term flight of capital out of the warring country, large-scale
emigration, and a deterioration of economic policy as measured
by a diagnostic assessment of the country's institutions and
policies by the World Bank.

A civil war reduces the average country's GDP per capita growth
by about one-third in the long run, with longer wars having a
stronger negative effect. However, researchers have also found
that there is potential for faster than normal economic growth in
the first decade following a conflict. Aid is also relatively more
effective in raising economic growth during a post-conflict period.
Valerie Cerra and Sweta Chaman Saxena confirmed that about
half of the loss in output is recouped within four years of a civil
war, but there still remains 3 percentage point cumulative loss
after a decade on average.

Politics

What determines the post-civil war nature of governance is an
active area of research with many competing theories. The type of

outcome (victory or negotiated settlement), the winning party (rebels or government), the type of war (ethnic or nonethnic), and the presence of outside groups (like the UN) are the most frequently cited determinates in the research. Mehmet Gurses and T. David Mason find that negotiated settlements are the most likely to result in democracy after a war, but other researchers, including myself, find no relationship.

Like me, Reyko Huang finds that a rebel victory is more likely to lead to democratic politics than a government victory, with Huang making a further distinction that the rebels must have been heavily reliant on the population during the conflict. This seems to indicate that victorious rebels gained both the capacity and the legitimacy needed to govern, which might explain why their victories tend to be longer lived. But this does not mean that democracy and the advancement of human rights directly follow. One need only look at the Taliban in Afghanistan. Generally speaking, political scientists Michael Doyle and Nicholas Sambanis find that democracy is more likely following nonethnic conflicts, but others have found no impact.

There is some research that points to UN peacekeeping operations as a positive influence on post-war democracy, but there is still much disagreement on this point. In a literature review of the effectiveness of peacekeeping operations, Jessica Di Salvatore and Andrea Ruggeri, for example, find that the effect United Nations peacekeeping operations have on democracy is not adequately studied, and have not been proven to help develop post-conflict democracy across cases.

Lastly, a deeper problem with the association of "good" outcomes with the nature of a civil war termination has to do, again, with how culturally determined the definition of "good" gets to be. For states in the OECD, which tend to be wealthy and are by definition democratic, the problem is that not all peoples agree that democracy is the best form of government; nor do they consider women's rights to be human rights. In places these are not considered ideal—and most of these countries cluster in the global South—it then becomes difficult to say that, for example, a rebel victory results in something "better" than an incumbent victory. The same might be said of economic development, which in many places counts as an aggregate measure of a country's wealth, rather than an adequate measure of its citizens' well-being. In many places, including within the OECD (the United States being a prime example), an increasing GDP only means that the rich are getting richer, not that the country as a whole is better off. So again, if it were the case that one sort of civil war ending correlated strongly with post-war economic development as compared to others, it doesn't follow that the post-war well-being of citizens has improved as a result.

Recurrence

One of the most important consequences of a civil war is the increased likelihood of another civil war. Fifty percent of all civil war conflict episodes between 1989 and 2018 recurred. This pattern is the result of the "conflict trap," the idea that conflict increases the likelihood of future conflict, whether via the conflict itself or the because the root causes of conflict go unaddressed. While most researchers recognize the conflict trap in one way or another, there are many diverging views about what causes it. The determinants of conflict recurrence researchers study most are peace agreements and their enforcement, the nature of the previous conflict, and the post-conflict government and economy.

Since World War II, the World Bank and others have documented that civil wars are far less likely to end in peace agreements than international wars. However, this might not necessarily be a problem as some researchers think negotiated settlements are less stable than outright victories by one side given the strong tendency to renege on commitments once the conflict ends.

One possible solution to this enforcement problem is the presence of peacekeeping operations, usually implemented by the UN. One study found that peace agreements supported by peacekeeping operations were substantially more likely to hold. Other researchers argue that the contents of the negotiated settlement are what matters most for sustaining peace. Caroline A. Hartzell and Matthew Hoddie identify four areas of power sharing—political, military, territorial, and economic—and show that the more areas addressed by the agreement the stronger the post-war peace. They argue that after a disastrous mass killing of ethnic Tutsi by Huti in Burundi in 1993, political and military power sharing prevented a subsequent civil war, where civil war or mass killing were previously considered very likely. The inclusion of electoral participation provisions has also been shown to promote peace, as these provisions are also easier to monitor and enforce. Lastly, while many researchers find evidence that agreements specifying the demobilization, disarmament, and reintegration (DDR) of rebel forces tend to lead to more durable peace, one convincing study showed that when DDR leads to a breakdown between rebel leaders and their lower-ranking field officers conflict tends to re-emerge.

One key factor of the previous conflict that tends to determine risk of war recurrence is who wins the war. Many researchers conclude that a rebel victory is more likely to lead to durable peace than a negotiated settlement or government victory. Again, one of reasons might be due to the legitimacy and popular support that a rebel organization gained in defeating the incumbent government.

The duration and intensity of the conflict also have an impact on likelihood of future conflict.

Håvard Hegre and his coauthors found that the deadlier and longer a conflict is the more likely that country will experience another conflict in the near future. Conversely, there is some evidence to suggest that a longer previous war makes subsequent war less likely since combatants are tired of fighting. So, clearly disagreements remain with mixed findings in the literature. The difficulty in finding clear and consistent findings can even be seen in the complex history of war recurrence in one country. For instance, Liberia finally seems to be at peace, but this is only after more than a decade of war and over a dozen negotiated settlements.

Governmental policy and its post-war economic performance are the final key drivers of civil war recurrence. First, there is a tendency for government military spending to remain high even after the conflict ends, and studies show that this spending is highly correlated with the risk of future conflict. High defense spending in South Sudan after its independence is a prime example. Economic development, however, tends to work in the opposite direction, with greater economic growth tightly linked to a lower likelihood of conflict recurrence. Combined with the fact that economies have the potential to grow quickly after war, outside investment in the post-war economy is a crucial step to stabilizing the peace.

Most researchers find that democracy after the war has no effect on the likelihood of recurrence, but a broader measure of "good governance" that includes other factors like bureaucratic quality and corruption does appear highly correlated with the likelihood of future conflict. The broad conclusion is that good governance rapidly reduces probability of recurrence after the initial war, but bad governance undermines peace in the long term.

UN interventions and peacekeeping operations

Beyond helping prevent civil war from recuring, external interventions in general, and UN peacekeeping operations in particular, are associated with higher levels of peace. UN peacekeeping operations are consistently shown to reduce civilian and battle deaths and help contain the violence geographically.

The larger the mission, the stronger these effects seem to be, and having a "robust" mandate is better than having a small or narrowly defined mission. Multiple researchers find that peacekeeping and mediation efforts by outside forces reduce the level of killing and help resolve conflict with a more inclusive settlement. Doyle and Sambanis also find that the positive influence of peacekeepers is stronger when they are introduced earlier in the conflict.

But peacekeeping operations have not been proven to help promote democracy, encourage human rights, or increase the rate of economic development. In this sense, UN intervention has been shown to encourage "negative peace" (or the lack of violence) but does not necessarily play a peacebuilding role. This would be the equivalent of effectively treating the symptoms of a chronic illness but never curing the patient. Without engaging these underlying causes (and armed forces are ideal for halting violence and preventing its recurrence but not trained or equipped to address underlying causes), once peacekeepers leave, the chances of civil war re-ignition rise.

Finally, providing external support in the form of money or weapons solely to the rebels, as might be the case in proxy warfare, undermines the bargaining process and tends to prolong the conflict.

Partitions after ethnic wars

There is a debate in the literature about whether partitioning a state between ethnic groups is an effective way to ensure stable peace following an ethnic civil war. Chaim Kaufman most famously claimed that partition is necessary because hyper-nationalist rhetoric and atrocities harden ethnic identities during war, and intermingled populations create security dilemmas, intensify violence, and prevent de-escalation. Others argue that partition will only lead to further fragmentation of the state.

Researchers also disagree as to whether ethnicity plays a role in the duration and intensity of civil war. One study found that when the war type changed from nonethnic to ethnic the probability of peace within two years was substantially decreased whereas others find no differences between ethnic or ideological conflicts in terms of the probability of peace. Nils-Christian Bormann and his coauthors attempt to resolve these contradictions. They posit that the inclusivity of power-sharing institutions after an ethnic war changes the likelihood of which groups engage in future conflict, while also noting that more regional autonomy (a form of partitioning) has no impact on the likelihood of future civil conflict.

Conclusion

The consequences of civil wars go far beyond who wins and who loses. Regardless of how the war ends, there are very real immediate and long-lasting human, economic, political, and security. But what if a civil war never occurred? This is an impossible question to answer definitively as we never truly know what the counterfactual would be. But the statistics and previous research give us a good idea of the average effects.

In a typical country if a civil war never occurred, many more people would be alive, more civilians would still live in their original homes, the population would be healthier and better educated, jobs would be more lucrative and incomes would be higher, and the country would face a lower risk of a civil war. At the same time, it took a bloody civil war for Cambodia to free itself from the genocidal Khmer Rouge regime. For many, the war could have been worth the cost.

Of course, a lot depends on how the civil war ends (which I outline in more detail in a later chapter). If the war's resolution can create the conditions for good governance, then the country's long-term prospects and the livelihoods of its citizens will be greatly improved. If poor governance follows, then renewed troubles are likely. But it is not only the citizens of the country at war that determine its fate. Outsiders can affect the war's outcome and the country's trajectory. It is first worth considering how external states are affected by civil wars before turning to how outsiders get involved—in both good and bad ways.

Chapter 5
Transnational effects

A large wave of Rwandan refugees began crossing the border into Zaire (now the Democratic Republic of Congo (DRC)) in April 1994. Trying to escape the genocide and civil war in Rwanda, millions of fellow Rwandans soon followed in search of safety in neighboring countries. Massive waves of civilians poured into large refugee camps just over the border in Zaire. But the perpetrators of the genocide soon infiltrated the camps, politicizing and militarizing the area. Armed groups used the camps as bases of operations and recruitment, complicating the end of the Rwandan civil war and destabilizing eastern Zaire.

Relying on support from the Mobuto regime in Kinshasa, the groups mounted cross-border attacks into Rwanda, angering the new Rwandan government in Kigali. These actions contributed to start of the first Congo War, sometimes dubbed Africa's first world war, when Rwandan troops invaded the country and helped topple Mobuto. The new Congolese government (the country's name was changed from Zaire to the DRC after Mobuto's fall) would not last long, however, and another civil war soon broke out, attracting foreign military forces from at least six other countries. Millions of people were displaced by the fighting, and it was one of the deadliest conflicts the world had seen since World War II.

In the end, the genocide and civil war in Rwanda affected the entire region, and the destabilizing effects linger on today. Rwanda is not unique in this respect. The Vietnam War extended beyond the sovereign borders of Vietnam and directly influenced the conflicts in neighboring Laos and Cambodia. Colombian rebel groups relied on international drug trafficking to fund operations during the conflict in Colombia, attracting the attention of the United States in its nominal war on drugs. Elements of the Taliban and al-Qaeda relocated to Pakistan to elude NATO and new Afghan government forces in the war in Afghanistan, much the same way as the Mujahideen relied on Pakistan as a safe haven in its war against the Soviet-backed Afghan government several decades earlier. And the civil war in Syria destabilized the whole region, with large numbers of refugees even reaching Europe and shifting local politics to the political far right.

Clearly the effects of wars spill over international borders, widening the war and exporting instability. What happens in a civil war country does not stay in the country. Rebel groups, especially those with pockets of ethnic compatriots across the border, often operate in relative safe havens, trying to evade government forces back home. Refugees flee outside the country to save their own lives. And conflicts and the movements of militants and displaced persons also assist illicit trade in arms, drugs, people, and other illegal goods.

These transnational journeys of armed groups, people, and goods can sow chaos in neighboring states. Diseases spread, economies suffer, local tensions mount, and the risk of another war rises. Concerns about security threats and the dispersion of conflict encourage outsiders—foreign fighters, non-state armed groups, and other countries—to get involved in the war to protect their own interests. Local conflicts can quickly become internationalized wars. In this chapter, we will examine the transnational effects of civil wars. We will first discuss what typically crosses international borders and why, before assessing how this affects the economies and security of other countries. It will soon become apparent that civil wars are rarely, if ever purely, internal.

What crosses borders?

Once a civil war begins, people and things begin to move. Often this movement transcends national borders. Rebels move for strategic reasons to increase their chances of winning the war against the regime. To escape death and destruction, to survive, non-combatants—the elderly, children—move. And it is not just the immiserated who escape decamp from their homes; opportunists take advantage of the movements and gaps in government control and the rule of law to set up illicit trade networks for arming warring parties, exporting drugs, diamonds, or other goods around the world, and transporting humans, often against their will, across borders. These activities can fund the continuation of the very war that is enabling the illegal trade in the first place and complicate its ending.

Armed groups

At least initially, rebel groups are typically weaker than their government opponents. Their forces are smaller. Their weapons are inferior. And their territory is much more confined. In fact, rebel groups often operate in government-controlled areas under constant threat of detection and detainment, or worse. This makes it difficult to evade capture, stockpile weapons, train forces, mount conventional military operations, establish consistent funding streams, or gain ground against government forces.

Rebel groups seek out places with minimum government control to establish operations. The strength of the regime's military tends to wane in the country's periphery or in mountainous or tough terrain far removed from the capital city, sometimes offering sanctuary for rebel groups to operate. But groups also congregate abroad. Research by Idean Salehyan finds that 55 percent of all rebel groups in civil wars since World War II utilize external sanctuaries outside of the country's sovereign borders, especially when neighboring states are weak and unable to clamp down on border crossings or easily eradicate foreign fighters on their territory. The winners of Rwanda's civil war in 1994 started from havens in Uganda, the Taliban and al-Qaeda sought relative safety across the border from Afghanistan in Pakistan, the Contra rebels escaped the grasp of Nicaraguan forces in Honduras and Costa Rica, the Tamil Tigers trained in India for their war against Sri Lanka's government, and Boko Haram operated outside of Nigeria, even mounting terrorist attacks in other countries in the Lake Chad Basin.

Government forces are frequently unwilling or unable to chase rebel groups beyond their own borders. Pursuing rebels abroad with enough force to win a military victory could easily anger the so-called host country, even if they do not want foreign rebel forces operating in the area. Transnational pursuit runs the risk of enflaming bilateral tensions and igniting an interstate war. Therefore, when rebel groups maintain access to foreign sanctuaries it makes it that much harder for their government opponents to conduct successful counterinsurgency campaigns and win a decisive victory in the civil war.

Rebel groups tend to relocate to areas abroad where there is weak government control, or they enjoy good relationships with local groups or governments. When their activities are not constantly under close government supervision and their members are not harassed or monitored, rebel leaders can set up the necessary organizational infrastructure to eventually wage a more evenly matched conventional war against their opponents. A foreign base of operations can be used to communicate and learn from likeminded groups, recruit new members, train existing forces, amass weapons, and develop steady sources of income by identifying foreign donors or establishing legal or illegal business operations. Thus, while their goals may remain focused on winning the civil war, their activities usually extend abroad.

Refugees

As violence spreads during a civil war, civilians can be targeted and are forcibly displaced from their homes (whether they are directly targeted or not), creating what Kelly Greenhill described as "weapons of mass migration." Fear of death, capture, or persecution force people to flee toward areas of relative stability in the country (leaving them as internally displaced persons) or abroad (making them refugees). The number of refugees is now higher than at any time since World War II and has nearly doubled in the past decade alone. To give you a sense of the scale, there are more refugees in the world today than the entire population of Australia. Half of Syria's prewar population was displaced with more than 6 million people crossing into other countries. Millions fled the wars in Afghanistan and Iraq. And more than a million refugees escaped violence in both South Sudan and Myanmar.

While a great deal of media attention is devoted to the refugees trying to make it all the way to Europe by boat or even cross the southern US border with Mexico after journeying across several other countries, most refugees do not travel that far. Most relocate to neighboring countries, which also tend to be developing countries with limited capacity to absorb the influx or means to provide for their survival. To help manage crises, UNHCR, or the UN refugee agency, was established in 1950 to support refugees, providing emergency assistance and trying to find long-term solutions. Along with nongovernmental relief organizations and local development groups, UNHCR works with governments to try to protect innocent civilians.

We tend to think of refugees living in giant camps just across the border, like those housing Rwandan refugees in Zaire. Then after the civil war ends, we assume refugees will instantly be able to return to their former homes. But this rarely happens. Refugees often relocate to urban areas outside of preplanned tented camps in search of jobs or rely on personal contacts. Individuals also tend to stay abroad for years or even decades. Civil wars often persist year after year and there may be no home to return to even after the fighting subsides. And if they return too soon or to an unwelcoming environment, they could reawaken hostilities and any reduction in violence could be lost. One need only think of the plight of Palestinian Arabs, who were forced to flee their homes and olive groves in 1948 in what they call the Nakba, or Catastrophe. Millions were displaced and never permitted to return. One of their most persistent demands in the decades-long peace processes that followed has been the right to return home.

While refugees are innocent victims of war, they still pose challenges for host states. The numbers of refugees in some host states are staggering. Pakistan, Iran, Turkey, Colombia, Uganda, and other countries have all hosted more than a million refugees at some point, and refugees have made up significant proportions of the populations of smaller states like Lebanon and Jordan. Sympathetic feelings among local populations soon give way to feelings of competition and resentment. Refugees are quickly seen as rivals for jobs and government attention. Their arrival can skew domestic balances among ethnic and religious groups and exacerbate prior tensions in the area, especially when they are ethnic kin with groups that are already marginalized. While fears that refugees will spread infectious diseases are overblown and demonstrate the type of backlash immigrants often endure, human displacement does complicate efforts to contain outbreaks, as was seen during the Ebola epidemics in West Africa and Democratic Republic of Congo.

Arms, drugs, and trafficking

The flight of people and groups from famine, extreme weather events, government repression, organized crime, and war makes illegal trade less risky and more profitable. As governments lose control over their own territory and borders during civil war, warring parties, criminal groups, and individuals can take advantage by importing desired materials and exporting goods to far-flung international markets. There is often a proliferation of small arms and uncontrolled weapons trade. Lootable resources like oil and gold, or resources that are relatively easy to obtain and move like drugs and diamonds, are often used to fund insurgencies. And human trafficking can increase with traffickers exploiting the vulnerability of displaced populations or setting up smuggling operations so people can pay for assistance in cross-border travel.

Demand for weapons and specifically small arms increases during civil wars. Rebel groups need weapons to wage war against more established government military forces, and individuals and criminal outfits may rely on their own arms to protect their property, groups, and operations. While data on trade in small arms and light weapons are limited, especially illegal and unreported trade among non-state armed groups, there are perhaps a billion small arms circulating around the world, international trade is worth billions of dollars, and small arms can be moved relatively easily across porous conflict borders. One of the most telling recent examples was during and after Libya's civil war in 2011 when large quantities of small arms were trafficked in the Sahel and used by rebel groups in Mali's civil war in 2012.

Warring parties regularly rely on illicit trade for funding. Drugs, diamonds, and other lootable resources are relatively easy for small groups to produce or acquire and then export to international black markets. Without government interference, civil wars often provide the space necessary for illegal trade to thrive. Some 90 percent of the world's opium comes out of the war in Afghanistan, and Colombia has accounted for a similar percentage of the world's cocaine supply at times during its own civil conflict despite significant US efforts to eradicate crops in its war on drugs. Conflict diamonds, commonly known as blood diamonds, helped fuel civil wars in Angola, Sierra Leone, Ivory Coast, and Democratic Republic of Congo, with significant portions of diamonds sold around the world coming out of conflicts. The problem became so severe that a multilateral trade regime called the Kimberly Process was established in the early 2000s to reduce the flow of conflict diamonds.

Human trafficking is a horrific part of war with vulnerable populations often used as child soldiers, forced laborers, or sexual slaves within the country or transported abroad to benefit their captors financially. The income generated from trafficking can fund belligerents and prolong fighting. Displaced populations are easy to exploit and can be deceived by false promises of safe transport, education, or international jobs before being forced into slave labor. Exact numbers are impossible to come by, but the International Labor Organization, Walk Free Foundation, and International Organization for Migration have estimated that there are more than 40 million victims of modern slavery, including forced labor and forced marriage. Boko Haram garnered global headlines when it abducted hundreds of girls from a school in Nigeria in 2014, the Islamic State sexually exploited Yazidi women in Iraq, and Rohingya populations were trafficked to Malaysia, Thailand, and other countries during the conflict and ethnic cleansing in Myanmar.

How are other states affected?

Armed groups, people, and illicit goods coming out of civil war countries affect the societies, political situations, economies, and security of other states. Economists tend to refer to these costs as "negative externalities," and they are significant. Data indicate that civil wars tend to disrupt the political balance in host countries, harm regional economies, and increase the risk of war within and among neighbors. But it is not always clear exactly why this is the case. We have limited knowledge in how these dynamics affect other states, although it is apparent that civil wars reduce economic growth and harm the stability and security of other states, particularly those that are already susceptible to political violence and internal conflict.

Economic consequences

Countries undergoing civil war suffer economically, and the same holds true for neighboring countries. The economic growth rates of neighboring countries tend to decrease, and research by James C. Murdoch and Todd Sandler indicates that the harm can stretch hundreds of miles beyond the conflict country's borders. War increases uncertainty and decreases regional investment, often disrupting regional or global supply chains. Murdoch and Sandler estimate that a civil war in another country within 500 miles negatively affects the economy (by almost a third as much as if the country was suffering its own civil war), but over time countries can insulate themselves from further economic harm.

Civil Wars

Refugees are often scapegoated as burdens on a host country's economy, but the actual impact is complicated. Short-term, a mass influx of refugees can be costly, though the arrival of international organizations and major nonprofit outfits and sudden surge in relief and development money can often provide a boost to local communities. But long-term, the economic gains to a host country's economy far outstrip initial costs. The effects, however, vary from location to location and from person to person. Individuals who work in the same fields as new immigrants may see increases in job competition and reductions to their incomes. And the emergence of black-market economies for illegal goods can cause detrimental effects as long as the conflict continues and the illicit trade endures.

Security implications

Civil wars tend to cluster in regions whether due to conflict contagion or because countries within the same geographic area often share similar risk factors for war. If there is a civil war in a neighboring country, it increases the risk of the country experiencing its own civil war. In fact, this turns out to be one of the best and most consistent predictors of civil war. Civil war neighborhoods are bad places to be. One of the reasons offered for conflict contagion is that large waves of refugees destabilize host states by straining the resources of governments, spreading arms and ideologies, disrupting the ethnic or religious balance in the area, and aggravating pre-existing local tensions. When non-state armed groups rely on external sanctuaries they can also offer knowhow and resources to domestic opposition groups, increasing the chance that they will choose to initiate their own rebellion. This is effectively what happened when an international and domestic coalition of anti-Taliban forces, led by the United States after the September 11, 2001, attacks defeated the Afghan Taliban in 2001, forcing most of them to flee to Pakistan, which had significant spillover effects in the tribal areas along the Afghanistan-Pakistan border. The area became a sanctuary for militant groups, including the Taliban, which regrouped and relaunched attacks in Afghanistan and against targets in Pakistan.

But civil wars do not only increase the risk of other civil wars, they also increase international tensions and the likelihood of interstate wars. If a country's military decides to pursue its domestic opponents across borders, it will anger the country hosting the rebel groups. And fears of conflict contagion or other negative effects associated with civil wars in other states may help compel a neighboring state to intervene in the civil war on behalf of rebel groups, aggravating bilateral tensions. Multiple conflicts within a region can quickly become linked by the participation of the same countries or connected non-state armed groups, and civil wars can also turn into proxy wars between external rivals.

Long-term problems

Security risks for other countries do not disappear when a civil war ends. There are potentially long-term consequences of civil wars that are often neglected by academic scholars. Refugees or other emigrants who have yet to return home may not support the outcome of the war or the future direction of country. Powerful diaspora communities can influence the prospects of sustained peace, and they can either choose to return, which could reignite a conflict, or back future uprisings. Wars may never truly be over until the issue of long-term refugees is resolved. The right of return continues to bedevil any potential resolution to Israeli-Palestinian conflict more than 70 years after Palestinian refugees were displaced in 1948.

Foreign fighters who traveled to participate in the civil war may also return to their country of origin or a new location. Political scientist David Malet estimates that the numbers of foreign fighters participating in civil wars is rising. Tens of thousands of foreign fighters have participated in conflicts in Muslim areas since 1980, with more than 40,000 fighters from more than 100 countries joining the Islamic State alone during its wars in Syria and Iraq. These fighters potentially pose a risk to whichever country they choose to relocate to next. Even if the civil war ends, its negative effects may live on abroad.

Conclusion

Civil wars are never purely domestic affairs. Armed groups, refugees, and illicit materials cross borders, affecting the economies and security of other states. The negative transnational effects of civil wars widen conflicts, increase local and international tensions, and cause external states to get involved, altering the dynamics of the war.

All of these "negative externalities" highlight an often-underappreciated feature of contemporary, as opposed to classical, civil wars: communications technology. Just as the railroad shifted power from maritime powers such as Britain to continental powers such as Germany, the container ship shifted power from the United States and Britain to countries with cheaper labor pools. It also made it cheaper to move large numbers of people from war-torn to OECD countries. The Internet has also made it cheaper and less risky to flee, because access to reliable travel routes and diaspora support networks make it easier for refugees to avoid interception and to find jobs or sanctuary abroad.

We will turn to external involvement in the next chapter. Rwanda's civil war can be linked to the subsequent civil wars in Zaire and later Democratic Republic of Congo. After the genocide, the new Rwandan government initially intervened to defeat elements of the former regime and overthrow Mobuto, and then decided to intervene again a few years later when its original ally, Laurent Kabila, turned on his international backers. These military interventions continue to affect the region's stability decades later.

Chapter 6
External involvement

Civil wars have transnational effects—few of which are welcomed by neighboring countries or the international community. A common phenomenon in the modern history of civil wars has been third-party intervention, or an outside state or international body deciding to get involved in another state's internal conflict. But not all interventions are due to civil war externalities like drug smuggling or refugees. States, the United Nations, or regional organizations may choose to intervene for geopolitical gain, humanitarian concerns, or economic benefit, along with many other reasons. In this chapter, I outline the rationales for intervention and how they affect civil wars.

When first considering "intervention," the minds of many students of history or foreign policy will immediately think of military intervention, recalling images of the UN's famous blue helmets for peacekeepers or Russian jets flying through the Syrian sky. In fact, this is far from the only type of intervention—though it is usually the most consequential, and always the most hotly debated.

Definitions of "intervention" vary, but the consensus is that intervention is the use of an outside actor's resources to affect the course of a civil conflict. This definition is wide, and intentionally so. An intervention can be any sort of expenditure of resources to impact a civil war's outcome. During the heyday of British imperialism, for example, an effective intervention could be as simple as the gift of a Rolls Royce automobile, along with a paved track on which to drive it. James N. Rosenau's definition, also widely accepted, proposes that intervention is both "convention-breaking" and "authority-oriented in nature." Convention-breaking means that intervention deviates from the normal and expected pattern of behavior between the intervening state and the target state, while "authority-oriented in nature" signals that the intervention intends to alter the authority or political structure in the target state in some way.

Types of interventions

Consider a playground metaphor again in which a bully is picking on a weaker classmate. If the two begin to fight, and a fellow student steps directly in between them to break it up, is this an intervention? We would likely say yes, because the student was using her resource (herself) to get involved in a conflict (the playground fight) with the aim of changing its outcome (to end the fighting). But, let us consider instead, a teacher witnesses the scene and breaks up the dispute with verbal instructions—is this an intervention? Again, the answer is yes. The teacher did not have to physically get involved in the fight but used her resource (authority) to halt the violence.

While this metaphor is obviously simplistic, it does have some parallels to the complicated world of international politics in which civil wars play out. Though the bulk of attention is paid to military interventions (and in fact, the bulk of this chapter is devoted to them), like the student's action, diplomatic interventions, as symbolized by the teacher's response, also fall under the general definition of "foreign intervention" in civil wars. A third type of intervention is economic, which is also common.

Diplomatic interventions

Diplomatic interventions—at minimum—attempt to manipulate the preferences of the warring factions, and thus alter the conflict outcome. The most common form of diplomatic intervention is conflict mediation, with a third party attempting to use negotiations to broker a peace, or at least a cessation of hostilities, between two parties. Mediation can be defined as a non-coercive, non-violent, and ultimately non-binding form of intervention.

Mediation is unique from other forms of intervention because it is entirely voluntary. The parties in question consent to the third-party brokering, and must agree on the format, location, and range of issues to be discussed prior to beginning. Perhaps the most helpful outcome of mediation is improved communication and trust between the two parties. The third-party advisor works to foster communication and information sharing and other confidence-building measures between the warring sides so that a reasonable settlement can be reached. Ultimately, successful mediation depends on an understanding of what each party desires and fears; and bluntly, this is neither obvious nor trivial. Many mediations fail as a result of a deeply held assumption that "everyone fears death above all else," which turns out not to be true. As Jake Bercovitch's research notes, a mix of carrots and sticks is essential to mediation effectiveness, but what counts as a "carrot" and a "stick" will often vary from conflict to conflict. A credible threat to kill is not enough.

In a famous scene from Richard Attenborough's epic biopic *Gandhi* (1983), the title character is addressing a crowd of angry young men, incensed by a new British law which enables any British policeman to enter an Indian home without a warrant and declares that only Christian marriages are recognized. In the audience are British officials, who become increasingly nervous as the anger of the young men grows. But Gandhi is insistent that India resist the British without violence:

> They may torture my body, break my bones, even kill me. Then, they will have my dead body, *not* my obedience.

It is no exaggeration to say that this idea, steadfastly represented by Gandhi's and Nehru's skillful and committed leadership, definitively ended British rule in India, which had depended for years on the "stick" of a credible threat to imprison, torture, or kill any who opposed British rule.

Civil war mediations can be undertaken by both state and non-state organizations. The United Nations is the most common mediator of civil wars, conducting 89 diplomatic interventions in 22 conflicts from 1947 to 2002 in all five regions of the world. UN diplomacy has played important roles in mediating the Arab-Israeli conflict in the immediate aftermath of World War II and more recently in Sierra Leone, Liberia, and Sudan. Perhaps surprisingly, in this same period, the Catholic Church also played an important peacebuilding role, especially in Latin America, serving as a mediator in the civil wars in El Salvador, Guatemala, and Mexico. In terms of state actors, the United States is the most frequent mediator. A notable example is its pivotal role in negotiating the Dayton Peace Accords, ending the bloody civil war in Bosnia in 1995.

While mediations are the most common, other forms of diplomatic interventions include recalling ambassadors, international forums, and arbitration. These efforts usually intend to focus the eyes of the international community on the conflict, to pressure the two sides in a conflict to reach a settlement, or to indicate international disapproval. They may not be as forceful in material terms as other measures, but they directly engage identity and legitimacy, are relatively cheap for the intervening state, and often cause some kind of reaction.

Economic interventions

Like diplomatic and military interventions, economic interventions most often attempt to change the behavior of civil war parties by making it more financially difficult for one or both sides to fight. Outside actors can send cash, weapons, and logistical supplies (like trucks, for example) in the hope that their favored combatant can convert that aid into victory. This was what the United States did in the Chinese Civil War following World War II. In that case, the economic intervention (like the later one in Indochina) was not sufficient to result in a Chinese Nationalist or French victory, respectively.

Today, the main form of economic intervention we see is sanctions. To limit their financial resources and access to necessary goods and materials, economic sanctions usually involve blocking trade with a government or group in question or freezing their international assets. All types of goods can be sanctioned, with the most obvious being weapons or materials that can directly be used by militaries. For example, during Liberia's civil war, the country was banned from buying arms on the global market—and restrictions were only lifted 13 years after the conflict formally ended.

However, sanctions often transcend the military realm and can also include basic foodstuffs and other consumer goods. After the civil war broke out in Syria in 2012, the prices of vegetables in Damascus rose so sharply that salad became a luxury good. The effect on salad prices illustrates the debatable efficacy of sanctions—they are intended to target militant organizations and oppressive leaders, but often end up hurting civilians the most by depriving them of even the most basic necessities such as food and medicine.

In addition to blocking parties from buying goods on the international market, sanctions can also limit a warring party's ability to sell to other countries. The Democratic Republic of the Congo was long restricted from selling gold on the international market, and Liberia only recently was allowed to export its timber and diamonds—two highly fought over minerals in its civil war—to foreign customers. Sanctions can also include ending foreign and humanitarian aid programs, blocking the use of currency on foreign markets, denying access to international financial institutions, and limiting an individual's ability to invest or move around the globe.

Like other types of intervention, sanctions can be undertaken unilaterally or multilaterally—though they are rarely effective unless they are a multilateral effort. The UN is the most popular coordinator of sanctions efforts, and sanctions are often coordinated and authorized by the Security Council.

Military interventions

Like its diplomatic and economic counterparts, military interventions in civil wars can take several different forms. One of the most common is a state deploying its own troops in another state's civil war—for example, the United States sending US Marines to Da Nang, Vietnam, in 1965. The United States had committed itself to support South Vietnam in its civil war with the North's Democratic Republic of Vietnam, and steadily escalated the quality and quantity of it support until 1968, when its leaders recognized that there was no practical way to win that war. Intervening parties usually choose to support one side over the other, as they may hope they can gain from their victory. Russia's intervention in the Syrian civil war in 2015 is another example, with Moscow using airstrikes and sending ground troops to support its political ally, Syrian leader Bashar al-Assad, against anti-government rebels.

Military interventions can also be undertaken without supporting either side. UN peacekeeping operations involve sending UN forces into conflict zones, but they are often mandated by the

United Nations against taking sides in the dispute. Peacekeepers are certainly a military presence—and are authorized to use force in self-defense—but usually cannot choose to support one faction over another, and instead "keep the peace" between the warring parties.

In modern war, however, a country or organization does not even need to send troops to intervene militarily in a civil war. Airstrikes, covert operations, or remote assistance have become a popular form of military intervention and present significantly less risk for the intervenor. The North Atlantic Treaty Organization (NATO) famously used airstrikes to alter the outcome of the Bosnian civil war in 1995 and again to prevent the massacre of civilians in Libya in 2011.

A party can also intervene by providing military intelligence or surveillance. American support of French military intervention in Mali in 2013—which was certainly a military intervention of the first sort, as France sent 1,700 troops into the fight—could qualify because the United States provided intelligence, transportation for ground troops, and aerial refueling to France, but declined to become directly involved. Another indirect type of military intervention is the simple provision of military equipment or logistical support intended to aid a military effort. An intervenor can decide to send weapons or other military resources to arm one side, hoping to shift the military balance in the war.

Lastly, deploying military advisors to a conflict also counts as an intervention. States or organizations may choose to send military trainers, strategists, or planners to help bolster the preparation and planning of one side in a civil war. This has become a common strategy for the United States, whose elite military forces are on the ground in places like Syria, Mali, Niger, and Somalia, training foreign troops and building the capabilities of their local partners.

Why do third parties intervene?

By now, you may be asking yourself why would a state intervene in another's affairs? We know that all countries have limited resources, and often many demands placed on those resources. Plus, a foundational principle of international law is the total sovereignty of the state and a reciprocal respect of that sovereignty by other states. Why then, considering resource constraints and international law, would a state choose to involve itself in another's civil war?

The answers vary, but no matter what the rationale, an intervening state will always undertake a cost-benefit analysis. In any intervention, it is likely that a third-party intervener determined that despite resource constraints, the action—whether for economic, reputational, or security reasons would result in a net benefit. Moreover, no state wants to intervene when the prospects of victory or an end to the war is elusive, resulting in a potentially long and expensive—both materially and reputationally—engagement.

Identity

Identity ties, be they ethnic, linguistic or religious, can motivate a third party to act in another's civil war. The outside actor may be a part of the same identity group as the non-state party and feel obligated to respond as a result of this shared affinity. As Steve Saideman illustrates in his book *The Ties that Divide*, countries with ethnic or religious kin might take sides thereby contributing to the conflict or impeding the ability to come to a resolution of the war. In this way, identity acts as a sort of alliance, forging a feeling of solidarity between the third party and the oppressed or rebelling group. This makes sense. Even in everyday life, you are more likely to take action to help a family member before you would a total stranger.

The impact of ethnicity was seen in Serbia's decision to attack Croatia and Bosnia during the Yugoslav civil war, as some argue that Croatian threats against its internal Serb minority were a large part of Serbia's rationale for its 1991 attack. We can also see ethnicity's impact in the war in Nagorno-Karabakh between Azerbaijan and Armenia, which started as a series of clashes in 1988 nd escalated into full-scale war by 1991, continuing until a ceasefire in 1994. So, what first started as a conflict between a group of Armenian separatists and the Azeri state soon became an interstate conflict, as Armenia intervened on behalf of its embattled ethnic kin and a full-fledged war broke out—the reverberations of which are still being felt today.

Economic considerations

Economic interests can also drive interventions, though interventions solely to protect economic interests have become far less common in the twenty-first century. There are several reasons why a third party may choose to intervene for economic reasons, with the first being plunder. Plunder has played a large role in African civil wars, which are often partly caused by conflicts over resources. In the Congo Wars from 1996 to 2003, for example, Zimbabwe capitalized on existing domestic instability and used military force to extract lucrative stones and metals. Former rebel leader in the Democratic Republic of the Congo, Wamba Dia Wamba, also asserts that both Uganda and Rwanda looted the country's precious stones, gold, and timber in the conflict, though both states deny it.

Third parties may also intervene in a civil war on behalf of a trading partner if they feel that the conflict will lead to a loss of trade. In the separatist war in the Casamance region of Senegal from 1984 to 2003, evidence indicates that Gambia was motivated to intervene by its illegal timber trade with Senegal, even switching sides in the conflict to support the rebels after the government lost control over the timber market.

Lastly, a state may also intervene to protect its own material interests in the country. In the nineteenth and twentieth centuries especially, industries in powerful countries—like the United Kingdom and United States—could count on their home countries to intervene to protect them. For example, in the 1920s the United States sent warships and over 2,000 Marines to Nicaragua to protect American mining interests during the Nicaraguan civil war and ended up in a five-year-long guerilla war against a rebel group led by Augusto Sandino (later the inspiration for the Sandinista insurgency against Nicaragua's Somoza dictatorship in 1979).

Transnational externalities

Before the age of the container ship and the mobile smart phone, critics of intervention to halt civil wars could rely on most of the harmful effects remaining local. Today, however, negative externalities—harmful transnational effects are less and less tethered to the territories on which these wars are fought. This can incentivize third-party interventions. As we discussed earlier, civil wars can have spillover effects into surrounding states, with refugees, illegal drug smuggling, cross-border raids, and other harmful effects spreading to other countries. A state that borders the country experiencing civil war may choose to intervene in the conflict simply to prevent these harmful effects from destabilizing its own internal affairs. Turkey's intervention in the Syrian Civil War from 2016 onward is a good example. Turkey's intervention was largely driven by security concerns about the establishment of an autonomous Kurdish region along its southern border, which could potentially embolden Kurdish separatist movements within Turkey itself. Additionally, Turkey aimed to manage the flow of refugees and maintain a buffer zone to prevent further influxes of displaced persons into its territory, thus addressing both internal security and humanitarian concerns.

In addition to physical externalities, there are ideological consequences of civil wars that may have transnational effects. For example, a state may fear that whatever ideology caused the conflict in its neighbor could have a "contagion effect," in which the belief system could spread across its borders and take root there—further destabilizing its regime—and may choose to intervene before that happens. States may also fear a demonstration effect, where a neighboring rebellion, especially a successful one, could inspire those within their territory to rebel. Relatedly, civil war could spread infectious diseases and other public health concerns, lowering standards of living and causing discontent with the host government.

Geopolitical implications

The possible strategic or geopolitical implications of a civil war may also encourage a third-party intervention. These are the most common type of intervention, which makes sense, because despite frequent claims to the contrary, states remain fundamentally driven by their own self-interest. States want to protect their homelands from threats emanating from a warring territory or wish to gain a stronger position because of an intervention. A state may choose to intervene if it feels the conflict is weakening a traditional rival or enemy, and therefore a total defeat by its opponents would strengthen its relative status. The United States chose to arm rebels fighting in the 1979 Afghan War against the Soviet occupiers for this exact reason—they sought to weaken the USSR by prolonging the war. There is some evidence that the Russian Federation may have returned the favor after 2001 in the same area for the same reason.

Keeping in mind that "stability" is a prerequisite of trade and economic growth, a third party might also intervene because of fears over regional stability. This is true for both states and international organizations. The United Nations is the most common face of an intervention, and often justifies its actions as a defense of regional peace and security. Other third-party regional organizations, like the African Union and the Economic Community of West African States (ECOWAS), have also intervened in the name of regional stability, seeking to end a conflict before it expands beyond its borders or sets off other tensions in the region.

Outsiders intervened diplomatically, economically, and militarily in the civil war in Mali to maintain regional stability. The intervention was first diplomatic and economic, with the UN Security Council leading the effort to sanction Mali and cut off support to Islamic jihadists and Tuareg rebels in the northern parts of the country. As the conflict progressed, ECOWAS prepared to take the lead and intervene militarily in Mali, but it was beaten to the punch by France, who intervened unilaterally in January 2013, following a request from the Malian government. As French President Hollande said after the fact, the conflict in Mali was important because "the disintegration of Mali's territorial integrity and constitutional government imperiled the stability of West Africa." Though French troops stayed on the ground in Mali after the initial intervention, the United Nations Multidimensional Integrated Stabilization Mission in Mali (MINUSMA) soon took over. Each of these different interventions were undertaken in the name of regional stability.

Humanitarian concerns

In addition to regional stabilization, humanitarian concerns commonly motivate UN interventions. If there is knowledge of widespread civilian harm resulting from civil war, including instances of war-caused starvation, torture, or mass killings and genocide, a third party may intervene to protect the innocent civilians, believing it to be a moral duty to help them. However, serious problems remain with this idea.

The original Geneva Conventions of 1949, for example, only bound and limited *states* and armed groups that could reasonably be claimed to be acting on behalf of states in an armed conflict. Within states, government and rebel forces were not bound by international law to obligations of restraint in their use of violence in pursuit of victory. This in part explains two things. First, the savagery of insurgency and counterinsurgency wars fought during the period of decolonization extending mainly from the end of World War II to the late 1960s. Second, the onset of the Cold War, with its ever-present fear of a species-ending global thermonuclear exchange between the USSR and United States, made the savagery of civil wars seem less important and, when seen and considered, a necessary price to pay to prevent World War III.

But efforts to protect non-combatants in civil wars moved forward anyway in what became Additional Protocols to the Geneva Conventions of 1949 (these came into effect in 1977). The aim of these protocols was to extend the protections and obligations of the original Conventions to belligerents in civil wars, but in particular to civil wars in which the insurgents were struggling against a "racist" incumbent. Of the three targets of Protocols I and II that were to be obligated and protected—Israel in the Occupied Territories, Portugal in Angola and Mozambique, and South Africa's Apartheid government—only Israel is still engaged in armed conflict. Israel did not sign or ratify the Additional Protocols, though the Palestinian [Arab] Authority ultimately did (its chief difficulty was that in 1974 it was not a state; so only after it was granted "observer status" in international law could its representatives accede to the Protocols).

Debates over the Additional Protocols represented the vanguard of a broader movement, which only accelerated after the Cold War's end, toward making respect for state sovereignty conditional on restraint when non-combatants were involved. The atrocities in Somalia, Haiti, Bosnia, Rwanda, and Kosovo solidified the idea that the international community needed to step in during extreme situations when civilians were in serious danger.

But to gain traction, the practice of humanitarian interventions needed to reconcile tensions between humanitarian assistance and state sovereignty. Some argued that the two concepts were fundamentally incompatible, even deeming NGOs aiding vulnerable populations during internal conflicts as a violation of sovereignty. Others believed that states had an obligation to let international assistance reach its intended targets. Since even humanitarian aid was heavily contested, the idea of an actor using military force to protect another state's vulnerable populations proved a difficult hurdle to overcome.

This tension came to define disputes over the Responsibility to Protect (R2P), a doctrine introduced by the UN in 2001. After the international community's failures to stop genocides in Rwanda and Bosnia in the mid-1990s, it became clear that a stronger mechanism was needed to mobilize and legitimize interventions by the United Nations and other regional actors. International attention and a sense of moral obligation was not enough. As Javier Perez de Cuellar, who served as UN secretary general at the end of the Cold War, stated, "We are clearly witnessing what is probably an irresistible shift in public attitudes toward the belief that the defense of the oppressed in the name of morality should prevail over frontiers and legal documents." When originally introduced by the International Commission on Intervention and State Sovereignty, R2P required six criteria to be met before an intervention: just cause, failure of the host state to act, right intention, last resort, proportional means, and reasonable prospects of success.

However, by the time R2P became an official legal document itself in 2005, many of these conditions were dropped. Still, the international community remained hesitant to act, in part because each state could imagine a condition under which another state might level a claim that it was violating the human rights of some portion of its citizens, thus not only permitting, but *obligating* outside intervention. If one state's oppression was another state's "social stability," R2P would be reduced to cultural—foreign or majority—constructions of what constituted a "human" right. It would be a fig leaf for cultural imperialism, favoring those with powerful militaries and economies or, bluntly, the global North and "West."

A second motivation to hesitation was intervention failures. The number of military interventions that have proven successful, regardless of the motivation, since World War II have been increasingly rare. The costly failures of UN humanitarian interventions in Somalia in 1993 and Rwanda in 1994 remain painful cases in point.

To skirt the sovereignty contradiction, R2P reframed sovereignty as an obligation that a state has toward its citizens—a state's sovereignty required it to be accountable to those under its jurisdiction. If a state failed to do so, the international community was then required to act. As Gareth Evans and Mohamed Sahnoun wrote in their 2002 *Foreign Affairs* piece defending the concept, "It is now commonly acknowledged that sovereignty implies a dual responsibility: externally, to respect the sovereignty of other states, and internally, to respect the dignity and basic rights of all the people within the state . . . sovereignty as responsibility has become the minimum content of good international citizenship." However, military intervention is still seen as a last resort of R2P with economic and diplomatic efforts coming first, while military action is reserved for extreme and intractable cases.

As adopted unanimously by the UN General Assembly in 2005, R2P is activated only in cases of a genocide, war crimes, crimes against humanity and ethnic cleansing. But as observed above, we are left with the question of who gets to decide? What happens if a finding of ethnic cleansing or genocide is made against a powerful state armed with nuclear weapons, such as the United States or Peoples Republic of China? Moreover, we have already seen the doctrine abused by fiat: in 2014 (and again in 2022), the Russian Federation claimed, without evidence, that Ukrainian "Nazis" were practicing ethnic cleansing against ethnic Russians in Eastern Ukraine, thus obligating the Russian Federation to intervene militarily in Ukraine's Donbass region. The Russian Federation under the leadership of Vladimir Putin has made a regular practice of issuing Russian passports to ethnic Russians in bordering states. What is to prevent the Russian Federation from a series of military interventions to "protect" Russian citizens in places such as Lithuania or Moldova, and justify such interventions under a theory of R2P?

Peacekeeping

Peacekeeping operations are probably the first thing that comes to mind when considering military interventions in civil wars, and they are certainly among the UN's most visible operations. Put simply, peacekeeping operations are military operations authorized by the United Nations that include military personnel from several countries and are usually tasked with "keeping the peace." Peacekeeping preceded R2P, with the earliest instances occurring in the early days of the United Nations in the 1940s. Peacekeeping missions are typically triggered by humanitarian concerns but can also be a response to a wide range of other catalysts, such as regional instability.

The United Nations is by far the largest initiator and steward of peacekeeping operations. Peacekeeping activities were originally meant to be guided by three principles, all of which have since shifted over time: consent of the host state, impartiality with respect to the warring parties, and use of force only in self-defense. Such principles remained intact during early peacekeeping activities, such as the monitoring of ceasefire agreements and demilitarized zones. However, as the nature of deployments shifted to messy and complex civil wars, these principles were altered to strengthen their mandates. Impartiality between warring parties became impartiality with respect to keeping the peace and upholding international law, while the use of force grew from only being authorized in self-defense to being authorized to uphold the mission's mandate. Peacekeeping activities also grew in scope. Peacekeepers helped create and maintain liberal democratic policies, temporarily served as state administrators, and enforced good governance. Critics of peacekeeping charge this wider scope and more expansive mandates have moved too far away from the original ideas of maintaining the peace during bloody conflicts or saving civilian populations from further harm. This criticism extends in part from the improper habit of associating peace with justice: in rare circumstances, war may be the only way to achieve justice and, in those cases, peacekeeping may perversely act to rescue repression and abuse.

With the Cold War's superpower gridlock no longer paralyzing the Security Council, UN-supported peacekeeping efforts grew significantly. Even before the concept of R2P was solidified, peacekeepers were deployed to intractable conflicts. In Bosnia peacekeepers made headlines for their inability to protect their own designated civilian "safe zones," hastening the shift to allowing more proactive uses of armed force.

In Somalia, UN troops monitored the first UN-brokered ceasefire in the country's civil war, though the fighting continued nonetheless. In more recent years, peacekeepers were deployed to places like Timor Leste, Central African Republic, and Mali to protect civilians and bring about national reconciliation. Civilian protection remains their most important purpose, and in the 2010s, over 90 percent of UN peacekeepers were explicitly mandated to protect civilians. Of the 139 major armed conflicts recorded from 1947 to 2013, 71 (or 51 percent) saw peace operations deployed as part of the international response. The numbers of recent decades are even higher, rising to 83 percent in the 2010s.

The UN, however, has no monopoly on peacekeeping. Regional organizations are increasingly engaged in peacekeeping operations, likely due to renewed tensions within the UN Security Council and the ambivalence of major powers about deploying their forces to another country's conflict. Regional bodies like the African Union, ECOWAS, and NATO have played pivotal roles in peacekeeping operations in the twenty-first century. The African Union, especially, has taken the lead in more recent conflicts and ranks second behind the UN as the most dominant actor in peacekeeping operations.

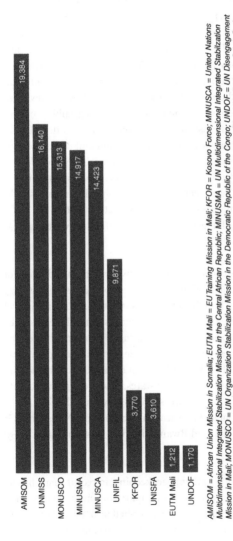

7. Largest multilateral peacekeeping operations as of December 31, 2021

AMISOM = African Union Mission in Somalia; EUTM Mali = EU Training Mission in Mali; KFOR = Kosovo Force; MINUSCA = United Nations Multidimensional Integrated Stabilization Mission in the Central African Republic; MINUSMA = UN Multidimensional Integrated Stabilization Mission in Mali; MONUSCO = UN Organization Stabilization Mission in the Democratic Republic of the Congo; UNDOF = UN Disengagement Observer Force; UNIFIL =UN Interim Force in Lebanon; UNISFA = UN Interim Security Force for Abyei; UNMISS = UN Mission in South Sudan.

Successes and complications of peacekeeping

Peacekeeping's track record is certainly mixed, but there are several promising statistics. Page Fortna reports that UN peace operations reduce the likelihood of armed conflict reigniting after a peace deal by 75–85 percent. Meanwhile, Jacob Kathman and Michelle Benson find that larger UN troop deployments can shorten the duration of civil wars, while Kyle Beardsley's research indicates that peacekeeping reduces the geographic scope of conflicts and the risk of violence spilling over borders. Lastly, the presence of peacekeepers reduces the amount of violence by deterring atrocities against civilians, lowering battlefield deaths, and helping to negotiate local settlements and intergroup settlements. Peacekeeping operations are also more cost-effective than other military operations. From 2013 to 2014, the UN peacekeeping budget was less than half of 1 percent of global military spending.

However, peacekeeping faces many challenges and does not often attain the level of success these broader statistics suggest. Peacekeeping both benefits and suffers from its multilateral nature. On the plus side, multilateralism implies cross-cultural consensus—a *shared* understanding of what constitutes crimes or emergencies. On the minus side, multilateral operations demand coordination and remain dependent on the political will and funding of UN member states. Few are eager to repeatedly risk their own armed forces. Disputes within the Security Council and lack of political will hampered UN responses in Bosnia, Somalia, and Rwanda in the 1990s, and again in Syria in 2012.

Many also contend that peacekeeping, in recent years, has not kept to its three principles of impartiality, minimal force, and consent. Instead, operations morphed into "peace enforcement" with the use of force allowed in more circumstances, but without proper authorization from the UN. In twenty-first century conflicts like Mali, Central African Republic, and Democratic Republic of the Congo, peacekeepers were deployed when there was "no peace to keep," often taking a side in the conflict, violating one of the core principles of peacekeeping. Peacekeepers can lose their protection under international humanitarian law if they violate impartiality, and in doing so often contradict their mandate.

Do military interventions in civil wars work?

Military interventions in civil wars are certainly common—of the 153 intrastate conflicts from 1945 to 2002, 100 experienced some sort of third-party intervention. Given their prevalence, it is worth asking whether they ever work.

The answer is predictably mixed. Christopher Linebarger and Andrew Enterline surveyed existing research to find that the efficacy of military interventions in civil wars is tied to which belligerent was being assisted. They write that external support is most consequential when it goes to insurgents, who are normally at a military disadvantage compared with the incumbent governments they are challenging. So if a third party intervenes to help the rebels, the rebels have an increased chance at winning. For example, Russia's interventions in Georgia and Moldova in the early 1990s helped separatist regions in each become effectively independent. However, when a third party intervenes on behalf of an incumbent government, the benefits are a lot less clear. Research is divided on the conditions that determine whether a government gains from outside support or not. Some find that pro-government interventions are effective only when rebel capabilities match or surpass those of the state, while others indicate that government-biased interventions are perceived by the local population as illegitimate, which then increases support for the rebels. Failed US efforts to prop up the government in Afghanistan after 2001 illustrate what can so often go wrong.

The impact of a third-party intervention on how long a civil war might last is also disputed. Civil wars, on average, last four times longer than interstate wars, a cost cited by many advocates of military intervention, who see it as a way to reduce human suffering by shortening the war. By contrast, other research finds that third-party intervention can prolong civil wars. Jeffrey Record has argued that third-party support for a weaker actor—generally insurgents as compared to incumbents, such as happened in the first and second Indochina wars—will most often prolong a conflict. A key question is whether prolonging a conflict by means of third-party intervention is a net benefit or cost. Certainly, there will be bloodshed either way, so the more difficult question facing researchers is not whether conflict duration is more costly in terms of killing, but whether the death and injury that invariably follows a prolonged conflict conduces to a legitimate outcome.

This problem is highlighted in a controversial article by Edward Luttwak, in which he argues that civil conflicts should be allowed to "burn out." Luttwak's argument only makes sense, however, if we accept his underlying premise: "might makes right." If true, that would mean any intervention on the part of a weaker actor would mean justice delayed. Third-party intervention can also make a final settlement more difficult to achieve, as third parties come to occupy yet another party at the civil war negotiating table whose preferences must be accommodated. Part of the Democratic Republic of Vietnam's (DRV) ongoing resentment of China stems from China's "betrayal" of the DRV after the decisive Vietnamese victory over the French at the Battle of Dien Bien Phu in 1954. The DRV expected its victory to result in the unification of Vietnam under its control, yet China's Zhou Enlai—a third-party intervenor on behalf of the DRV—negotiated a demilitarized zone (effectively keeping Vietnam divided into Communist North and Nationalist South) instead.

Chapter 7
Ending civil wars

Having experienced only 11 years of peace since gaining
independence from the British and after two rounds of civil war
from 1955–1972 and 1983–2005, over 99 percent of South
Sudanese people voted in favor of independence from Sudan in a
referendum in January 2011. Five months later, on June 9, the
Republic of South Sudan became the world's newest state. On that
day, UN Security Council resolution 1996 established the United
Nations Mission in the Republic of South Sudan (UNMISS) to
consolidate peace and security and assist in building state capacity
in the new country.

South Sudan's official independence represented the culmination
of the peace negotiations that finally ended war between the north
and south. The long-lasting second war began when Sudanese
president Jaafar Nimeiri made a series of controversial decisions
in 1983 that disregarded the 1972 negotiated settlement to the
previous Sudanese civil war, basically abrogating south Sudan's
hard-won autonomy. The Southern Peoples Liberation Army
(SPLA) responded by fighting against government-aligned forces
based in Khartoum. The government empowered tribal militias
(*murahaliin*) to undertake counterinsurgencies against the SPLA,
plundering, raping, and killing communities and causing famine.
Throughout the conflict, the two sides attempted several peace
negotiations and temporary ceasefires that ultimately failed.

Randolph Martin, a long-time observer and expert on Sudan, describes the second round of the civil war as the "perfect war" for the north. From the government's perspective, the war was for a long-time politically and economically self-sustaining, especially with access to vast oil wealth as a source of funding. The government also maintained diplomatic relations with almost every state, except the United States. The SPLA generally lacked incentives to pursue peace with the north, leading to the expanded goal of independence.

Eventually, the costs of the continued fighting on Sudan's government began to tell. Despite continued violence in the south, negotiators from the Intergovernmental Authority on Development (IGAD), a regional organization, along with the United Kingdom, United States, Norway, and Italy, facilitated peace negotiations in 2003 and 2004. In January 2005 the National Congress Party of the north and the SPLA signed a comprehensive peace agreement in Kenya. Prior to arranging for an independence referendum in 2011, the agreement outlined a power-sharing system and equitable oil revenue arrangement, ending the civil war. The proposed independence referendum would allow those in southern Sudan to decide whether they would continue under the deal's power-sharing agreement or become an independent state, which they ultimately chose in overwhelming numbers. Tragically, however, the new nation of South Sudan enjoyed only a brief period of peace before falling victim to yet another war between the new country's two main ethnic groups, the Dinka and the Nuer.

How civil wars end

Just as no two wars are alike, none of their endings are the same either. But broadly speaking, the main options are via a stalemate or ceasefire, a negotiated settlement, or with victory, either by the government or the rebel group. A stalemate or a ceasefire involves all sides laying down their arms and no agreement on the future make-up of the government. Negotiated settlements more often involve both sides laying down their arms and a shared agreement on and participation in a new government. Victories by either side are more intuitive—one side defeats the other thereby empowering the winning side to determine the future of the government and country.

So which ending is best? Does the type of ending affect the chances of a sustainable peace or a negative peace (the absence of war)? In answering these questions, three factors are often considered: (1) whether the civil war has "ended" or how likely it is to recur; (2) the expected costs (usually in terms of deaths) if the war re-ignites; and (3) the quality of the peace that results, or whether the post-war government is better or worse for the parties and their constituents.

If we assume, as many do (especially outside observers), that the greatest cost of war is death and destruction, it makes sense to think of negotiated settlements as the most desirable way to end a civil war. Finding a compromise that includes both sides, ending the killing and destruction as soon as possible, would be preferable to allowing one side to lose the war and be wiped out or forced to flee. That is most likely why this assumption prompts most research and policy-making to focus on negotiated deals, thinking this way of ending the violence reduces the costs to society and leads to better outcomes in terms of economic reconstruction, development, and political liberalization.

But despite this academic focus, 60 percent of all civil wars since 1940 ended by military victory and only 20 percent by negotiated settlement. (The remainder ended in ceasefires or stalemates.) Since 1990, however, civil wars are as likely to end by negotiated settlement (42 percent) as by military victory (39 percent). This trend could be attributed to the collapse of the Cold War's superpower rivalries and support for proxies in civil wars, and increased support for peace processes and settlements. Political scientists Lise Howard and Alexandra Stark argue that the norm of supporting negotiated settlements arose in a post-Cold War era dominated by American unipolarity and the liberal international order. Since the 2000s, the frequency of negotiated settlements has once again declined with the reemergence of great power rivalry, widespread fatigue with peacekeeping, and disenchantment with the failures of past settlements. In addition, as noted by Ivan Arreguin-Toft, the success rate of military interventions of all types has dropped steadily since 1950, a trend most recently exemplified by the 20-year-long failed effort by the United States to support Afghanistan's government against Taliban insurgents. Add to this the preoccupation of major powers, particularly the United States, Britain, and France with other concerns, such as religiously inspired terrorism. In the post-September 11 world, stabilization surpassed democratization as the main goal, leading to a decrease in all termination types, especially negotiated settlements that involve "terrorist" groups.

After their civil wars end, only 43 percent of countries make it a decade without another war. Roughly 50 percent of all civil wars between 1989 and 2018 recurred. Some scholars conclude that a conflict often leads to future conflicts, especially if the initial conflict was large. A comparison of all civil war outcomes since 1946 suggests that military victories are the most likely to lead to a decade of peace, with 61 percent classified as stable (using a decade of peace as a benchmark of post-conflict stability).

More specifically, military victories decrease the likelihood of war recurrence by 13 percent. Rebel victories seem to be the most stable, in part because they seem to have the best combination of legitimacy and capacity if you consider that rebels, typically the underdogs, have defeated a state that was supposed to have the monopoly on the legitimate use of violence. The ability to achieve such a feat is often a sign of popular support. Comparatively, only about half of peace agreements and ceasefires managed to sustain peace.

The nature of the civil war also affects the sustainability of peace, with ethnic conflict 36 percent less likely than nonethnic conflict to reach two years of peace. Michael Doyle and Nicholas Sambanis have shown that peacekeeping missions that support negotiated settlements can result in increased peace duration over time.

Military victories may lead to longer lasting peace for several reasons. At the deepest level, war is about how and why *people* fight. Access to plentiful and sophisticated weapons certainly remains a part of any explanation of victory or defeat, but in so many cases (e.g. the Communists in the Chinese Civil War, the Viet Minh in the Vietnamese Civil War, the Afghan Mujahideen in the Afghan Civil War), the side with few weapons, or old weapons, but unshakeable commitment won despite being outgunned. It may be then that military victory signals whose cause is relatively more just: if well-resourced incumbents cannot win quickly, perhaps it is because the insurgents' cause is more just. Military victories also affect two other causes of a prolonged outcome or a re-ignition. First, since a military victory means the winner disbands the loser's organization and surviving fighters, there's no one left to worry, as they might in a negotiated settlement, that once a peace treaty is signed, the more powerful side might change its mind and take advantage of the peace. Second, military victories reduce the number of post-war compromises that will need to be made to near zero, thus facilitating peacetime reconstruction and development.

The different termination types also differ in costs, affecting how we think about what is the "best policy." According to the 2012 Human Security Report, the death tolls from wars with negotiated settlements are about half of that of civil wars that end with a military victory. And despite the greater likelihood of conflict recurrence following negotiated agreements, those subsequent conflicts are less deadly than those that occur following a military victory, which is more likely to feature genocidal-type violence. Finally, the nature of the quality of peace—or positive peace determined by democracy and prosperity—varies based on the termination type as well. The wars ended by peace agreements tend to have been longer lasting wars than those that ended by military victory. Given that longer wars tend to have resulted in higher levels of destruction, disruption, and grievances, that will mean expanding on a negotiated peace once a war is over is likely to be relatively more challenging.

Of course, not all victories or settlements are the same. So, if we hope to anticipate the effects of different civil war settlements we need to unpack their substantive contents. For example, the nature of power sharing and other confidence-building measures along with the levels of aid and development provisions all affect the durability of a settlement. Keep in mind that power sharing's effectiveness comes with an underlying assumption that societies benefit from an environment in which opposing views can clash without any advocate having to fear retribution. This may explain why constitutional reform that creates space for debate, and checks and balances on authority, often correlates with a reduced likelihood of conflict re-ignition. Constitutional changes can signal a commitment to enforce a new political and social reality, reflecting bargains between formerly adversarial elites. Changes aimed at improving women's political rights after conflict requires the design process to explicitly consider and include women (more voices). Finally, third-party oversight of new constitutional processes can help to reduce uncertainty and insecurity in a new political order, deepening post-conflict peace.

Regardless of the way a civil war is ended, much depends on working out the plans for belligerents to demobilize. How will belligerents give up their arms (something which demands a *lot* of trust)? How will fighters return to civilian life? Will former insurgent fighters become part of a post-war state's regular military and, if so, how will that work? These plans, along with how faithfully they are implemented, all affect the sustainability and quality of peace as well, no matter the termination type.

Military integration, for example, involves committed participation from both of the formally warring parties in the new national military or police forces. Policies of inclusion, such as the ability for rebels to form political parties in government, help to ensure that peace holds by supporting a more comprehensive "buy-in" process.

Post-conflict stability tends to be the highest following military victories, regardless of whether the rebels demobilize or keep their militias. In terms of the quality of the peace, democracy often decreases over time, with democracy most likely to wane following negotiated settlements and least likely to diminish following a rebel victory. Successful rebel groups who enjoy significant material support from civilians are often less repressive than those rebel victors without support.

Successful settlements require a balance of the government's ability to benefit and harm relevant actors. Think "carrots" and "sticks." A government needs to be capable of providing benefits, like goods and services, while still being able to credibly threaten to support a post-war order with force. Stalemates and ceasefires rarely include sufficient harms or benefits, resulting in no significant peace. Negotiated settlements often assume the benefits (everyone agrees killing is costly), but not the harms (each side's sense that peace may betray justice and betray those who have already lost their lives in war), so peace can wither. Government victories mainly privilege the threat of harms over the provision of benefits, leading to enduring peace but likely under tyranny. Rebel victories are most likely to balance both benefits and harms, resulting in a greater probability of a more equitable society and possibility for peace.

The academic literature has largely ignored military victories. Scholars have tended to focus on different factors within negotiated settlements. Caroline Hartzell and Matthew Hoddie, emphasize institutions and electoral practices as keys to make negotiated settlements work better. Although there has been effort to consider disarmament, demobilization, and reintegration efforts in these settlements, few have closely evaluated security sector reform: how, after a peace agreement is signed, fighters will be reintegrated into the post-war state. Even among those who study state-building and democratization, there has been little mention of military or armed forces' roles in post-conflict institutions. Political scientists Ronald Krebs and Roy Licklider concede that "military integration often accompanies sustainable post-war peace," but political conditions influence local actors to embrace military integration, not the other way around. Therefore, they argue that military integration works best when the local communities are eager, while imposed integration can lead to the "integrated" army splitting back into warring factions.

In all, recent academic studies have focused too much on the external balances of power—the impact of third parties on outcomes—seeming to forget how important the internal balances of power are for influencing a country's future. This inattention to internal factors is made worse by the fact that external balances imposed by third parties are best at engaging the symptoms of an underlying conflict—fighting—temporarily freezing a conflict which then thaws and re-ignites when, inevitably, third parties depart.

The post-Cold War obsession with settlements has not always been predominant; previous research often stressed the consolidation of power as a prerequisite for reform. Older research also determined that rapid consolidation of the monopoly on the legitimate use of force is preferable than slower consolidation. Eminent scholars such as Charles Tilly, Samuel Huntington, and Barrington Moore focused on how violence and war are critical to building national unity and institutions, especially the military. As A. F. K. Organski described, "The colonizer who leaves gracefully before he is forced to leave does not act as beneficially as he imagines, for nations born of

Caesarian birth are weaker." Consequently, the older insights concluded that military victories are superior to negotiated settlements because they consolidate power by eliminating opponents, creating clear winners and losers, and allowing for rapid consolidation; whereas settlements and stalemates are more likely to result in a multipolar power structure with multiple winners and without a clear monopoly on the legitimate use of force. More centralized and unified states are more likely to have sustained peace.

Despite evidence that military victories may experience a more sustained peace, the resulting peace can vary greatly in duration. Partitions, where ethnic populations are separated into distinct territorial enclaves, are often considered to be one form of a rebel victory. Partitions can be *de facto*—physical, but not legal—or *de jure*—both physical and legally recognized, with both groups gaining autonomy or outright sovereignty and statehood. There are many arguments favoring partitions. When ethnic wars cause a fear-based security dilemma—a situation in which any group's efforts to increase its own security is necessarily assumed by all other groups to constitute a threat—different identity groups often consolidate for security purposes. Co-existence becomes impossible when violence destroys the foundation for cross-ethnic cooperation, as these identities are hardened by violent rhetoric.

Political scientist Chaim Kaufmann, in particular, argues that it is impossible to restore civil politics in civil war-torn multiethnic states, requiring the warring groups to be separated. He made this case strongly for former Yugoslavia. Ethnic civil wars are more likely to recur than nonethnic wars, and therefore when partitions divide populations into "defensible enclaves" the hope is that the new territorially distinct entities that are created can be sustained on the basis of mutual deterrence, with each side in a position to impose substantial costs on the other.

Partition, however, is an unpopular outcome to civil wars in both academic and policy circles. Some argue that partitions reward the use of violence by protesting minorities, leading to a moral hazard problem. This could then lead to future conflicts based on at least one side believing that further action is worth the risks. Partitions also have the potential, if not the high likelihood, to consolidate and harden historical and battlefield identities, which can privilege majorities and large minorities, leaving smaller groups aside with little recourse to gain representation and protection. Finally, partitioning is arguably never complete. Partitions can leave vulnerable populations in place and tend to lead to migration flows, potentially transforming a civil war into an interstate or regional conflict. Radha Kumar argued this pointing to the case of India and Pakistan. Ultimately, the partition debate remains unresolved, with proponents and opponents holding radically different views on the desirability, effectiveness, and necessity of partitions.

The precariousness of peace

Sudan is a recent example of partition and one seen as a last resort (again, after decades of bloodletting). The comprehensive peace agreement outlined how the violence between the north and south would end, whether the two sides decided to co-exist in one country or as two different countries.

Several factors contributed to this final deal. There was a demonstrated willingness by both sides to negotiate and an interest among the international community to find a resolution. A real settlement also required several confidence-building measures and incremental steps, involving a ceasefire in the Nuba mountains, "tranquility zones" to allow humanitarian relief, an end to aerial bombings of international organizations and civilian targets, and investigations into allegations of slavery. A peace agreement, the Machakos Protocol, served as an important first step in the formal process, outlining principles on governance, transitional processes and structures, and establishing principles of self-determination for South Sudan. Finally, the agreement included the key provision that the south would have the opportunity to vote for independence within five years.

It is important to evaluate the effectiveness of the deal since 2005. South Sudan President Salva Kiir Mayardit mainly relied on patronage and privilege to his fellow Dinka tribal members as his method of governance, favoring that over stability. As the South Sudanese leader since the end of the Sudanese civil war, he required access to resources to sustain power and client networks to safeguard his position. Without these resources, instability was inevitable, and his control was tenuous as non-Dinka leadership challenged him, notably Vice President Riek Machar, a member of the Nuer tribe. With attention focused on sustaining power, the security sector was mismanaged, unmonitored, and not held accountable.

At the time of independence, the SPLA had 745 generals, which was more than the combined US armed services. Such massive amounts of high-ranking officers inevitably demanded larger and larger budgets, regardless of the country's economic growth. Already in 2006, the military budget allocated over 80 percent to salaries and was overspent by 363 percent. To compound the problem, basic salaries doubled to $150 in 2006 and then increased to $220 by 2011. Furthermore, critical resources flowed to a narrow segment of South Sudan's society, rather than to the wellbeing of the general population, which faced starvation and war.

South Sudan both demonstrates the precariousness of peace after civil wars end and the complexity of implementing peace agreements and partition plans. Even after gaining independence supported by most of the local population and powerful international backers, the government in Juba still faces instability. Internal divisions, ethnic tensions, elite rivalries, and uncertainties over oil and other resources led to more civil war in South Sudan. The world's newest country did not enjoy a negative peace, instead suffering from increasing ethnic violence and rebellion with the SPLA seen as a Dinka tribal institution beholden to President Kiir. Any major conflict will likely spill over into other neighboring countries, further exacerbating regional tensions.

When considering how to end civil wars, third-party policy-makers need to consider what they value and prioritize the most. Stability, peace, democracy, or something else? They also, and in particular, need to consider what the civil war belligerents they hope to persuade value, prioritize, and, yes, fear the most.

Chapter 8
The past, present, and future of civil wars

Civil wars are the most common type of war today. No other form of large-scale political violence happens more often. Most civil wars in recent years are identity struggles—fights over ethnic or religious differences—and they almost always attract foreign fighters. They tend to be lengthy conflicts that are difficult to resolve, killing or displacing large numbers of civilians, delaying schooling for youthful populations, and destroying livelihoods, healthcare, critical infrastructure, and food sources. And even after the fighting stops, the road to recovery is long with the country facing higher odds of follow-on civil war.

If current trends persist, Syria will not be the last civil war that displaces millions and destabilizes an entire region for years to come. Azerbaijan, Democratic Republic of Congo, and Somalia will not be last countries to suffer recurring bouts of war. Myanmar and Ethiopia will not be the last wars pitting different identity groups against one another. South Sudan will not be the last country to win its independence in war. And Afghanistan, Libya, Mali, and Yemen will not be the last civil wars where external states fail to establish a stable and friendly government.

What we know

Once scholars and policy-makers turned their attention away from the interstate rivalries of the Cold War, and strategies for avoiding a global thermonuclear exchange, research on civil wars gained momentum. Academics compiled datasets, analyzed the numbers, researched the differences among types of wars, and identified important trends and patterns. We now have a shared knowledge base of the common risk factors for civil wars and how and why they start.

Civil wars are more likely in nondemocratic countries, but not necessarily autocratic states, that have already fought a civil war, are in a bad neighborhood, have large and youthful populations with gender inequalities, and are governed by an incompetent and corrupt leadership; one in which elites frequently attempt to maximize their rule or their wealth by driving a wedge between identity groups for their own gain. Religious tensions and ethnic divisions make a country particularly ripe for war, while unfulfilled expectations can also spur individuals and groups to rebel.

Once a civil war gets started, the consequences are likely to be severe. Civilians often suffer the most. Large numbers are killed or displaced from their homes, while education suffers, incomes drop, poverty increases, malnutrition soars, human rights abuses multiply, and social services disappear. While countries are sometimes able to quickly return to pre-war growth rates and rebuild infrastructure with international support, survivors who were directly affected by violence typically face long-term consequences. Girls and women who were raped will have painful choices to make about publicly acknowledging this form of torture or dealing with unwanted pregnancies. And the grievances all survivors must put away to again move forward become highly susceptible to manipulation by unscrupulous elites. These are only some of the reasons they will also face a higher likelihood that war will recur in their lifetimes.

While civil wars are fought within a state, their impact extends well beyond national borders. Armed groups, refugees, and illicit goods cross into neighboring states, and possibly bring along some of the problems from their home countries. Neighboring economies can suffer, and bordering states can be destabilized, increasing the risk of yet more war. The negative transnational effects sometimes compel external states to intervene, but these interventions are not always effective at ending the conflict and can even prolong the war by complicating possible resolutions.

In the 1990s we began to see more civil wars end in negotiated settlements. With funds from their Cold War backers drying up and enhanced international efforts to prevent humanitarian disasters, the chance of outright victories by either side started to fall. But this did not always turn out as well as proponents hoped, because unlike civil wars where there is a clear winner, particularly a rebel victor, the governments put together in negotiations were often disunited and weak and not sufficiently incentivized or equipped to manage unresolved grievances. Consistent external support and the deployment of international peacekeepers could help keep the peace, but war recurrence was a common problem.

Since 9/11, attention on negotiated settlements to civil wars and interest among the great powers for supporting sufficient peacekeeping forces have waned. Terrorism and great power conflict came to be seen as bigger concerns. The proportion of negotiated settlements has fallen accordingly, and altruistic military interventions to forestall humanitarian suffering are less likely.

What we need to learn

While we have a decent appreciation for where civil wars are possible, even likely, we are not very good at predicting when a new civil war will begin with any precision. There are simply too many factors to consider and insufficient knowledge about unique local settings, situations, and individuals to be as accurate as policy-makers would like. Deep knowledge of countries, cultures, and identities, supplemented by innovative and rigorous data analyses can help and should inform future scholarship.

One of the ways to improve our knowledge of when civil wars will occur is to research the connections among different forms of political violence and how conflicts escalate. When are protests and demonstrations likely to spread and turn violent, and when are small, localized, or non-state conflicts likely to turn into full-fledged civil wars? We also need to figure out how the domestic risk factors will be affected by the changing global balance of power, transnational armed groups including Salafi jihadists, and climate change. Threat multipliers like climate change will not have the same effect in every location, so conclusions will not be universal or simple. Moreover, although communication and technology have always been a part of civil wars, given the widespread availability of smart phones with Internet connectivity and the impact of social media platforms on how groups compare themselves to others, we still have a lot to learn about how these new technologies are likely to impact the onset, nature, spread, and ending of civil wars.

While we know much more today about how civil wars affect regional peace and stability, there is a lot more to learn. We do not know enough about the second-order effects of the transnational movements and communications of armed groups, foreign fighters, and refugees. And we also have more to learn about how the international community can best act to end ongoing civil wars, prevent recurring or new ones, and make any that do occur less destructive and less likely to affect neighbors, regions, and the broader interstate system. We need to innovate better tools and uses for economic and diplomatic interventions, while working to make military interventions rarer, but also, when absolutely necessary, adequately resourced, reasonably aimed, and above all discriminate. Taken together, these three traditional tools of statecraft can work to give belligerents (or imminent belligerents) incentives to stop the violence, while at the same time helping to address the deeper underlying issues that escalated to violence in the first place.

What the future holds

Today the world seems to be returning its attention to great power politics and the threat of war between states, but civil wars are not disappearing, and as always, create increased risks of great power military confrontations. The obvious candidates for new or renewed civil wars remain. Countries with a history of conflict or tensions among identity groups that are in insecure regions, like the Middle East, Sahel, or Central America's Northern Triangle, are at risk. Ukraine, at the heart of "stable" Europe, remains in conflict, and its neighbors east and west are nervous that they may be next. If these places are also straddled with corrupt and ineffective governments, and young and underemployed populations, the threat to peace and stability will be magnified.

But the risks are also growing for several countries that until quite recently seemed to have moved beyond civil war. As demographic changes lead to increasing tensions between identity groups and democracies backslide into anocracies, even wealthy countries in the West—including the United States—cannot dismiss the dangers. As I and others warned before and after the January 6, 2021, insurrection, no country is immune.

Civil wars are likely to continue to be the most common form of political violence in the world, so it is worth improving our ability to forecast when new wars are going to break out and implement policies that will pull countries back from the brink before a grudge becomes a fight, a fight becomes a war, and a war ends us all.

References

Chapter 1

Balcells, Laia. *Rivalry and Revenge: The Politics of Violence during Civil War*. Cambridge: Cambridge University Press, 2017.

Howard, Lise Morjé, and Alexandra Stark. "How Civil Wars End: The International System, Norms, and the Role of External Actors." *International Security* 42/3 (2018): 127–171.

Kalyvas, Stathis. *The Logic of Violence in Civil War*. New York: Cambridge University Press 2006.

Olson, Mancur. *The Logic of Collective Action: Public Goods and the Theory of Groups*. Cambridge, MA: Harvard University Press, 1971.

Toft, Monica Duffy. *The Geography of Ethnic Violence: Identity, Interests and the Indivisibility of Territory*. Princeton, NJ: Princeton University Press, 2003.

Wood, Elisabeth Jean. *Insurgent Collective Action and Civil War in El Salvador*. Cambridge: Cambridge University Press, 2003.

Chapter 2

Chenoweth, Erica, and Maria Stephan. *Why Civil Resistance Works*. New York: Columbia University Press, 2011.

Kaldor, Mary. *New and Old Wars: Organized Violence in a Global Era*. 3rd ed. Stanford, CA: Stanford University Press, 2012.

Kalyvas, Stathis. "'New' and 'Old' Civil Wars: A Valid Distinction?" *World Politics* 54/1 (2001): 99–118.

Leitenberg, Milton. *Deaths in Wars and Conflicts in the 20th Century.* 3rd ed. Occasional Paper, Cornell University Peace Studies Program no. 29. Ithaca, NY: Cornell University, 2006.

Sarkees, Meredith R., and Frank W. Wayman, eds. *Resort to War, 1816–2007.* Washington, DC: CQ Press, 2010.

Schock, Kurt. *Armed Insurrections: People Power Movements in Non-Democracies.* Minneapolis: University of Minnesota Press, 2005.

Sharp, Gene. *Social Power and Political Freedom.* Boston: Porter Sargent Publishers, 1980.

Small, Melvin, and J. David Singer. *Resort to Arms: International and Civil Wars, 1816–1980.* Beverly Hills: Sage Publications, 1982.

Chapter 3

Arjona, Ana. *Rebelocracy: Social Order in the Colombian Civil War.* Cambridge: Cambridge University Press, 2016.

Cederman, Lars Erik, Kristian Gleditsch, and Halvard Buhaug. *Inequality, Grievances, and Civil War.* West Nyack, NY: Cambridge University Press, 2013.

Collier, Paul, and Anke Hoeffler. "On Economic Causes of Civil War." *Oxford Economic Papers* 50/4 (1998): 563–573.

Connor, Walker. "A Nation Is a Nation, Is a State, Is an Ethnic Group, Is a...." *Ethnic and Racial Studies* 1/4 (1978): 377–400.

Fearon, James D., and David D. Laitin. "Ethnicity, Insurgency and Civil War." *American Political Science Review* 97/1 (2003): 75–90.

Kaplan, Oliver R. *Resisting War: How Communities Protect Themselves.* Cambridge: Cambridge University Press, 2017.

Kaufmann, Stuart. *Modern Hatreds: The Symbolic Politics of Ethnic War.* Ithaca, NY: Cornell University Press, 2001.

Mearsheimer, John J. *Conventional Deterrence.* Ithaca, NY: Cornell University Press, 1983.

Olson, Mancur. *The Logic of Collective Action: Public Goods and the Theory of Groups.* Cambridge, MA: Harvard University Press, 1971.

Sambanis, Nicholas. "Do Ethnic and Nonethnic Civil Wars Have the Same Causes? A Theoretical and Empirical Inquiry (Part I)." *Journal of Conflict Resolution* 45/3 (2001): 259–282.

Smith, Anthony. *The Ethnic Origins of Nations.* Oxford: Blackwell Publishers, 1986.

Snyder, Jack. *From Voting to Violence: Democratization and Nationalist Conflict.* New York: W. W. Norton & Co., 2000.

Staniland, Paul. *Networks of Rebellion: Explaining Insurgent Cohesion and Collapse*. Ithaca, NY: Cornell University Press, 2014.

Toft, Monica Duffy. *The Geography of Ethnic Violence: Identity, Interests, and the Indivisibility of Territory*. Princeton, NJ: Princeton University Press, 2003.

Toft, Monica Duffy. "Getting Religion?" *International Security* 31/4 (2007): 97–131.

Toft, Monica Duffy. "Identicide: How Demographic Shifts Can Rip a Country Apart." *The Conversation*, April 24, 2019.

Walter, Barbara F. *How Civil Wars Start and How to Stop Them*. New York: Crown Publishers, 2022.

Waltz, Kenneth. *Theory of International Politics*. Long Grove, IL: Waveland Press, 1979.

Weiner, Myron. "Bad Neighbors, Bad Neighborhoods: An Inquiry into the Causes of Refugee Flows." *International Security* 21/1 (1996): 5–42.

Williams, Paul. *War and Conflict in Africa*. Oxford: Polity Press, 2016.

Wood, Elisabeth Jean. *Insurgent Collective Action and Civil War in El Salvador*. Cambridge: Cambridge University Press, 2003.

Chapter 4

Balcells, Laia, and Stathis Kalyvas. "Does Warfare Matter? Severity, Duration, and Outcomes of Civil Wars." *Journal of Conflict Resolution* 58/8 (2014): 1390–1418.

Beardsley, Kyle, David Cunningham, and Peter B. White. "Mediation, Peacekeeping, and the Severity of Civil War." *Journal of Conflict Resolution* 63/7 (2019): 1682–1709.

Bell, S., and F. Huebler. *The Quantitative Impact of Conflict on Education*. Montreal: UNESCO Institute for Statistics, 2010.

Bormann, Nils, et al. "Power Sharing: Institutions, Behavior, and Peace." *American Journal of Political Science* 63/1 (2019): 84–100.

Buckland, Peter. *Reshaping the Future: Education and Post-Conflict Reconstruction*. Washington, DC: World Bank, 2004.

Cerra, Valeria, and Sweta C. Saxena. "Growth Dynamics: The Myth of Economic Recovery." *American Economic Review* 98/1 (2008): 439–457.

Collier, Paul, et al. *Breaking the Conflict Trap: Civil War and Development Policy*. Washington, DC: World Bank, 2003.

Daly, Sarah Zukerman. "The Dark Side of Power-Sharing: Middle Managers and Civil War Recurrence." *Comparative Politics* 46/3 (2014): 333–353.

Di Salvatore, Jessica, and Andrea Ruggeri. "Effectiveness of Peacekeeping Operations." *Oxford Research Encyclopedia* (online, 2017).

Doyle, Michael W., and Nicholas Sambanis. *Making War and Building Peace: United Nations Peace Operations.* Princeton, NJ: Princeton University Press, 2006.

Gurses, Mehmet, and T. David Mason. "Democracy Out of Anarchy: The Prospects for Post-Civil-War Democracy." *Social Science Quarterly* 89/2 (2008): 315–336.

Hartzell, Caroline A., and Matthew Hoddie. *Crafting Peace: Power-Sharing Institutions and the Negotiated Settlement of Civil Wars.* College Station: Penn State University Press, 2007.

Hegre, Håvard, Håvard Mokleiv Nygård, and Ranveig Flaten Ræder. "Evaluating the Scope and Intensity of the Conflict Trap: A Dynamic Simulation Approach." *Journal of Peace Research* 54/2 (2017): 243–261.

Huang, Reyko. *The Wartime Origins of Democratization: Civil War, Rebel Governance, and Political Regimes.* Cambridge: Cambridge University Press, 2016.

Jarland, Julie, et al. "How Should We Understand Patterns of Recurring Conflict?" *Conflict Trends* 3. Oslo: PRIO, 2020.

Kaufmann, Chaim. "Possible and Impossible Solutions to Ethnic Civil Wars." *International Security* 20/4 (1996): 136–175.

Kumar, Radha. "The Troubled History of Partition." *Foreign Affairs* 76/1 (1997): 22–34.

Matanock, Alia M. *Electing Peace: From Civil Conflict to Political Participation.* Cambridge: Cambridge University Press, 2017.

Murdoch, James C., and Todd Sandler. "Economic Growth, Civil Wars, and Spatial Spillovers." *Journal of Conflict Resolution* 46/1 (2002): 91–110.

Quinn, J. Michael, T. David Mason, and Mehmet Gurses. "Sustaining the Peace: Determinants of Civil War Recurrence." *International Interactions* 33/2 (2007): 167–193.

Shemyakina, Olga. "The Effect of Armed Conflict on Accumulation of Schooling: Results from Tajikistan." *Journal of Development Economics* 95/2 (2011): 186–200.

Toft, Monica Duffy. "Ending Civil Wars: A Case for Rebel Victory?" *International Security*, 34/4 (2010): 7–36.

UNICEF, "Education Disrupted: Impact of the Conflict on Children's Education in Yemen." United Nations Children's Fund, Yemen, 2021.

Weinstein, Jeremy. *Inside Rebellion: The Politics of Insurgent Violence.* Cambridge: Cambridge University Press, 2007.

Chapter 5

Barrett, Richard. "Beyond the Caliphate: Foreign Fighters and the Threat of Returnees." New York: The Soufan Center, 2017.

Buhaug, Halvard, and Kristian Gleditsch. "Contagion or Confusion? Why Conflicts Cluster in Space." *International Studies Quarterly* 52/2 (2008): 215–233.

Forsberg, Erica. "Transnational Dimensions of Civil Wars: Clustering, Contagion and Connectedness." In *What Do We Know about Civil Wars?*, ed. T. Mason and S. Mitchell, 75–90. Lanham, MD: Rowan and Littlefield, 2016.

Gleditsch, Kristian, Idean Salehyan, and Kenneth Schultz. "Fighting at Home, Fighting Abroad: How Civil Wars Lead to International Disputes." *Journal of Conflict Resolution* 52/4 (2008): 479–506.

Greenhill, Kelly M. *Weapons of Mass Migration: Forced Displacement, Coercion, and Foreign Policy.* Ithaca, NY: Cornell University Press, 2010.

Hegghammer, Thomas. "The Rise of Muslim Foreign Fighters: Islam and the Globalization of Jihad." *International Security* 35/3 (2010): 53–94.

Malet, David. *Foreign Fighters: Transnational Identity in Civil Conflicts.* New York: Oxford University Press, 2013.

Murdoch, James C., and Todd Sandler. "Civil Wars and Economic Growth: Spatial Dispersion." *American Journal of Political Science* 48/1 (2004): 138–151.

Salehyan, Idean. *Rebels without Borders: Transnational Insurgencies in World Politics.* Ithaca, NY: Cornell University Press, 2009.

Salehyan, Idean, and Kristian Gleditsch. "Refugees and the Spread of Civil War." *International Organization* 60/2 (2006): 335–366.

Chapter 6

Aydin, Aysegul. "Where Do States Go? Strategy in Civil War Intervention." *Conflict Management and Peace Science* 27/1 (2010): 47–66.

Beardsley, Kyle. "Peacekeeping and the Contagion of Armed Conflict." *Journal of Politics* 73/4 (2011): 1051–1064.

Committee on International Relations, House of Representatives, "Peacekeeping: Cost Comparison of Actual UN and Hypothetical U.S. Operations in Haiti." In Report to the Subcommittee on Oversight and Investigations. US Government Accountability Office, GAO-06-331, February 2006.

Evans, Gareth, and Mohamed Sahnoun. "The Responsibility to Protect." *Foreign Affairs* 81/6 (2002): 99–110.

Fortna, Virginia Page. *Does Peacekeeping Work? Shaping Belligerents' Choices after Civil War.* Princeton, NJ: Princeton University Press, 2008.

Kathman Jacob, and Michelle Benson. "Cut Short? United Nations Peacekeeping and Civil War Duration to Negotiated Settlements." *Journal of Conflict Resolution* 63/7 (2019): 1601–1629.

Koops, Joachim, ed. *The Oxford Handbook of United Nations Peacekeeping Operations.* 1st ed. New York: Oxford University Press, 2015.

Linebarger, Christopher, and Andrew Enterline. "Third Party Intervention and the Duration and Outcomes of Civil Wars." In *What Do We Know about Civil Wars?*, ed. T. Mason and S. Mitchell, 93–108. London: Rowman and Littlefield, 2016.

Luttwak, Edward. "Give War a Chance." *Foreign Affairs* 78/4 (1999): 36–44.

Saideman, Stephen M. *The Ties that Divide: Ethnic Politics, Foreign Policy, and International Conflict.* New York: Columbia University Press, 2001.

Regan, Patrick. *Civil Wars and Foreign Powers: Outside Intervention in Intrastate Conflict.* Ann Arbor: University of Michigan Press, 2000.

Rosenau, James N. "Intervention as a Scientific Concept." *Journal of Conflict Resolution* 13/2 (1969): 149–171.

Chapter 7

Arreguín-Toft, Ivan. *How the Weak Win Wars: A Theory of Asymmetric Conflict.* New York: Cambridge University Press, 2005.

Doyle, Michael W., and Nicholas Sambanis. *Making War and Building Peace: United Nations Peace Operations.* Princeton, NJ: Princeton University Press, 2006.

Fiedler, Charlotte. "Why Writing a New Constitution after Conflict can Contribute to Peace." Briefing Paper, No. 11/2019. Bonn: Deutsches Institut für Entwicklungspolitik (DIE), 2019.

Fortna, Virgina Page. *Does Peacekeeping Work? Shaping Belligerents' Choices after Civil War.* Princeton, NJ: Princeton University Press, 2008.

Hartzell, Caroline A., and Matthew Hoddie. *Crafting Peace: Power-Sharing Institutions and the Negotiated Settlement of Civil Wars.* College Station: Penn State University Press, 2007.

Hegre, Håvard, Håvard Mokleiv Nygård, and Ranveig Flaten Ræder. "Evaluating the Scope and Intensity of the Conflict Trap: A Dynamic Simulation Approach." *Journal of Peace Research* 54/2 (2017): 243–261.

Howard, Lise Morjé, and Alexandra Stark. "How Civil Wars End." *International Security* 42/3 (2017/18): 127–171.

Huang, Reyko. *The Wartime Origins of Democratization: Civil War, Rebel Governance, and Political Regimes.* Cambridge: Cambridge University Press, 2016.

Human Security Report Project. *Human Security Report 2012: Sexual Violence, Education, and War: Beyond the Mainstream Narrative.* Vancouver: Human Security Press, 2012.

Huntington, Samuel P. *Political Order in Changing Societies.* New Haven, CT: Yale University Press, 1968.

Jarland, Julie, et al. "How Should We Understand Patterns of Recurring Conflict?" *Conflict Trends* 3. Oslo: PRIO, 2020.

Joshi, Madhav, and Louise Olsson. "War Termination and Women's Political Rights." *Social Science Research* 94 (2020): 451–470.

Kaufmann, Chaim. "Possible and Impossible Solutions to Ethnic Civil Wars." *International Security* 20/4 (1996): 136–175.

Krebs, Ronald R., and Roy Licklider. "United They Fall" *International Security* 40/3 (2015/16): 93–138.

Martin, Randolph. "Sudan's Perfect War." *Foreign Affairs* 81/2 (2002): 111–127.

Moore, Barrington Jr. *Social Origins of Dictatorship and Democracy: Lord and Peasant in the Making of the Modern World.* Boston: Beacon Press, 1966.

Organski, A. J. P. *The Stages of Political Development.* New York: Knopf, 1965.

Quinn, J. Michael, T. David Mason, and Mehmet Gurses. "Sustaining the Peace: Determinants of Civil War Recurrence." *International Interactions* 33/2 (2007): 167–193.

Tilly, Charles. *From Mobilization to Revolution*. Reading, MA: Addison-Wesley Publishing Company 1978.

Toft, Monica Duffy. *Securing the Peace: The Durable Settlement of Civil Wars*. Princeton, NJ: Princeton University Press, 2009.

Walter, Barbara F. *Committing to Peace: The Successful Settlement of Civil Wars*. Princeton, NJ: Princeton University Press, 2002.

Chapter 8

Toft, Monica Duffy. "How Civil Wars Start." *Foreign Policy*, February 18, 2021.

Further reading

Arjona, Ana. *Rebelocracy: Social Order in the Colombian Civil War*. Cambridge: Cambridge University Press, 2016.

Arreguín-Toft, Ivan. *How the Weak Win Wars: A Theory of Asymmetric Conflict*. New York: Cambridge University Press, 2005.

Balcells, Laia. *Rivalry and Revenge: The Politics of Violence during Civil War*. Cambridge: Cambridge University Press, 2017.

Buckland, Peter. *Reshaping the Future: Education and Post-Conflict Reconstruction*. Washington, DC: World Bank, 2004.

Cederman, Lars Erik, Kristian Gleditsch, and Halvard Buhaug. *Inequality, Grievances, and Civil War*. West Nyack, NY: Cambridge University Press, 2013.

Chenoweth, Erica, and Maria Stephan. *Why Civil Resistance Works*. New York: Columbia University Press, 2011.

Collier, Paul, et al. *Breaking the Conflict Trap: Civil War and Development Policy*. Washington, DC: World Bank, 2003.

Doyle, Michael W., and Nicholas Sambanis. *Making War and Building Peace: United Nations Peace Operations*. Princeton, NJ: Princeton University Press, 2006.

Fortna, Virginia Page. *Does Peacekeeping Work? Shaping Belligerents' Choices after Civil War*. Princeton, NJ: Princeton University Press, 2008.

Greenhill, Kelly M. *Weapons of Mass Migration: Forced Displacement, Coercion, and Foreign Policy*. Ithaca, NY: Cornell University Press, 2010.

Hartzell, Caroline A., and Matthew Hoddie. *Crafting Peace: Power-Sharing Institutions and the Negotiated Settlement of Civil Wars*. College Station: Penn State University Press, 2007.

Huang, Reyko. *The Wartime Origins of Democratization: Civil War, Rebel Governance, and Political Regimes*. Cambridge: Cambridge University Press, 2016.

Human Security Report Project. *Human Security Report 2012: Sexual Violence, Education, and War: Beyond the Mainstream Narrative*. Vancouver: Human Security Press, 2012.

Huntington, Samuel P. *Political Order in Changing Societies*. New Haven, CT: Yale University Press, 1968.

Kaldor, Mary. *New and Old Wars: Organized Violence in a Global Era*. 3rd ed. Stanford: Stanford University Press, 2012.

Kalyvas, Stathis. *The Logic of Violence in Civil War*. New York: Cambridge University Press 2006.

Kaplan, Oliver Ross. *Resisting War: How Communities Protect Themselves*. Cambridge: Cambridge University Press, 2017.

Kaufmann, Stuart. *Modern Hatreds: The Symbolic Politics of Ethnic War*. Ithaca, NY: Cornell University Press, 2001.

Koops, Joachim A., ed. *The Oxford Handbook of United Nations Peacekeeping Operations*. 1st ed. New York: Oxford University Press, 2015.

Leitenberg, Milton. *Deaths in Wars and Conflicts in the 20th Century*. 3rd ed. Occasional Paper, Cornell University Peace Studies Program no. 29. Ithaca, NY: Cornell University, 2006.

Malet, David. *Foreign Fighters: Transnational Identity in Civil Conflicts*. New York: Oxford University Press, 2013.

Mason, T. David, and Sara McLaughlin Mitchell, eds. *What Do We Know about Civil Wars?*. Lanham, MD: Rowan and Littlefield, 2016.

Matanock, Alia M. *Electing Peace: From Civil Conflict to Political Participation*. Cambridge: Cambridge University Press, 2017.

Mearsheimer, John J. *Conventional Deterrence*. Ithaca, NY: Cornell University Press, 1983.

Moore, Barrington, Jr. *Social Origins of Dictatorship and Democracy: Lord and Peasant in the Making of the Modern World*. Boston: Beacon Press, 1966.

Olson, Mancur. *The Logic of Collective Action: Public Goods and the Theory of Groups*. Cambridge, MA: Harvard University Press, 1971.

Organski, A. J. P. *The Stages of Political Development*. New York: Knopf, 1965.

Regan, Paul. *Civil Wars and Foreign Powers: Outside Intervention in Intrastate Conflict*. Ann Arbor: University of Michigan Press, 2000.

Saideman, Stephen M. *The Ties That Divide: Ethnic Politics, Foreign Policy, and International Conflict*. New York: Columbia University Press, 2001.

Salehyan, Idean. *Rebels without Borders: Transnational Insurgencies in World Politics*. Ithaca, NY: Cornell University Press, 2009.

Sarkees, Meredith R., and Frank W. Wayman, eds. *Resort to War, 1816–2007*. Washington, DC: CQ Press, 2010.

Schock, Kurt. *Armed Insurrections: People Power Movements in Non-Democracies*. Minneapolis: University of Minnesota Press, 2005.

Sharp, Gene. *Social Power and Political Freedom*. Boston: Porter Sargent Publishers, 1980.

Small, Melvin, and J. David Singer. *Resort to Arms: International and Civil Wars, 1816–1980*. Beverly Hills: Sage Publications, 1982.

Smith, Anthony. *The Ethnic Origins of Nations*. Oxford: Blackwell Publishers, 1986.

Snyder, Jack. *From Voting to Violence: Democratization and Nationalist Conflict*. New York: W. W. Norton & Co., 2000.

Staniland, Paul. *Networks of Rebellion: Explaining Insurgent Cohesion and Collapse*. Ithaca: Cornell University Press, 2014.

Tilly, Charles. *From Mobilization to Revolution*. Reading, MA: Addison-Wesley Publishing Company, 1978.

Toft, Monica Duffy. *The Geography of Ethnic Violence: Identity, Interests and the Indivisibility of Territory*. Princeton, NJ: Princeton University Press, 2003.

Toft, Monica Duffy. *Securing the Peace: The Durable Settlement of Civil Wars*. Princeton, NJ: Princeton University Press, 2009.

Toft, Monica Duffy. Daniel Philpott, and Timothy Samuel Shah. *God's Century: Resurgent Religion and Global Politics*. New York: W. W. Norton & Company, 2011.

Walter, Barbara F. *Committing to Peace: The Successful Settlement of Civil Wars*. Princeton, NJ: Princeton University Press, 2002.

Walter, Barbara F. *How Civil Wars Start and How to Stop Them*. New York: Crown Publishers, 2022.

Waltz, Kenneth. *Theory of International Politics*. Long Grove, IL: Waveland Press, 1979.

Weinstein, Jeremy. *Inside Rebellion: The Politics of Insurgent Violence*. Cambridge: Cambridge University Press, 2007.

Williams, Paul. *War and Conflict in Africa*. Oxford: Polity Press, 2016.

Wood, Elisabeth Jean. *Insurgent Collective Action and Civil War in El Salvador*. Cambridge: Cambridge University Press, 2003.

Index

Figures are indicated by an italic *f* following the paragraph number.

Index

Civil Wars